Student Solutions Guide

Laurel Carpenter ▪ Sherry Biggers

to accompany

Calculus Concepts
An Applied Approach to the Mathematics of Change

Fourth Edition

LaTorre ▪ Kenelly ▪ Reed ▪ Carpenter ▪ Harris ▪ Biggers

Houghton Mifflin Company **Boston** **New York**

Publisher: Richard Stratton
Senior Sponsoring Editor: Molly Taylor
Senior Marketing Manager: Jennifer Jones
Senior Development Editor: Maria Morelli
Editorial Assistant: Joanna Carter-O'Connell
Marketing Associate: Mary Legere
Editorial Associate: Andrew Lipsett
New Title Project Manager: Susan Peltier

We gratefully acknowledge the contributions of **Jennifer LaVare**, **Jason Martin**, **Carrie Green**, and **Judith McKnew** for their contributions to the answers and solutions for the fourth edition of *Calculus Concepts*.

Printed in the U.S.A.

ISBN-10: 0-618-78986-3
ISBN-13: 978-0-618-78986-3

3456789–CRS–11 10 09 08

Contents

Chapter 1
Ingredients of Change: Functions and Models

Section 1.1 Models and Functions

1. **a.** Input description: weight of letter
   Output description: first-class domestic postage
   Input variable: w
   Output variable: $R(w)$
   Input units: ounces
   Output units: cents

   b. R is a function of w because a letter of one weight cannot have two different domestic first-class postage amount.

   c.

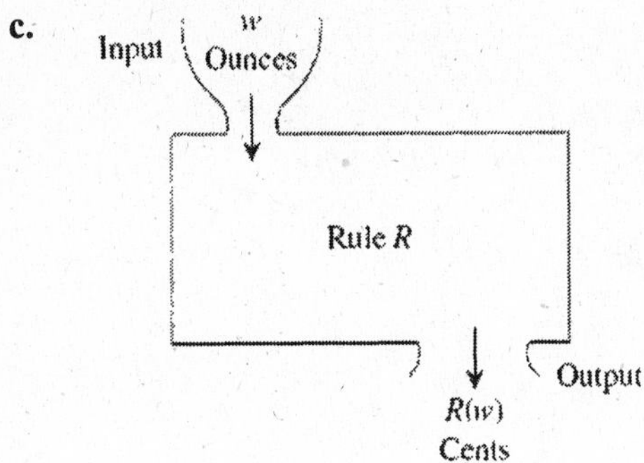

3. **a.** Input description: day of the week
   Output description: amount spent on lunch
   Input variable: m
   Output variable: $A(m)$
   Input units: none
   Output units: dollars

   b. A is not a function of m unless you always spend the same amount on lunch every Monday, the same amount every Tuesday, etc., or unless the input is the days in only 1 week.

   c.

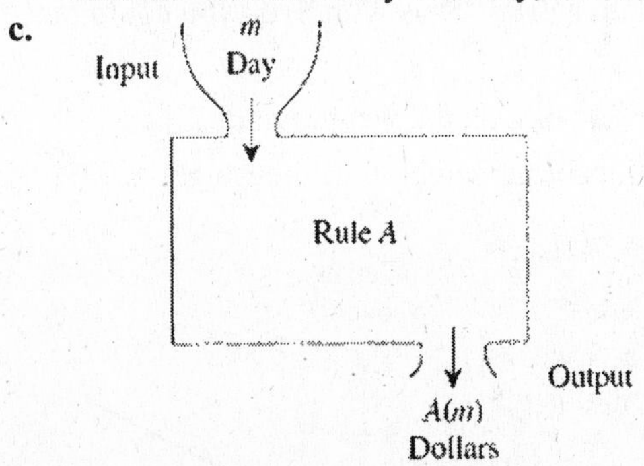

5. The table represents a function because each input (year) corresponds to only one output (number of iPods sold).

7. The table represents a function because each input (depth of dive) corresponds to only one output (maximum dive time).

9. Graphs *b* and *c* are functions. Graph *a* is not a function because vertical lines cutting through the circle touch it at two points.

11. **a.** *P*(Honolulu, HI) = 295

b. *P*(Providence, RI) = 137.8

c. *P*(Portland, OR) = 170.1

13. **a.** In 1988 cotton exports had a value of $1,975,000,000.

b. In 1992 cotton exports had a value of $1,999,000,000.

15. **a.**

Cost (dollars)

144, 126, 108, 90, 72, 54, 36, 18

x CDs

1 2 3 4 5 6 7 8 9 10

b. The cost of 6 CDs is the cost of 5 CDs at $18 each, plus 1 free CD:

(5 CDs)($18 per CD) = $90.

c. You could buy 2 CDs for $36. Because $36/$18 per CD = 2 CDs.

d. The graph shows 6 CDs for $90 and 7 CDs for $108. Thus with $100 you could buy 6 CDs.

17. At birth the baby weighed 7 pounds, so (0, 7) is a point on the weight graph.
After 3 days $\left(\frac{3}{7}\text{ week}\right)$ the baby has lost 7% of its birth weight, thus the weight is 93% of the birth weight: 0.93(7) = 6.51 pounds, and $\left(\frac{3}{7}, 6.51\right)$ is a point on the graph. At 1 week, the weight is again 7 pounds; at 2 weeks, the weight is 7.5 pounds; at 3 weeks, the weight is 8 pounds; and at 4 weeks, the weight is 8.5 pounds. Thus we have the points (1, 7), (2, 7.5), (3, 8), (4, 8.5). Plotting these points and connecting them with line segments results in the following graph:

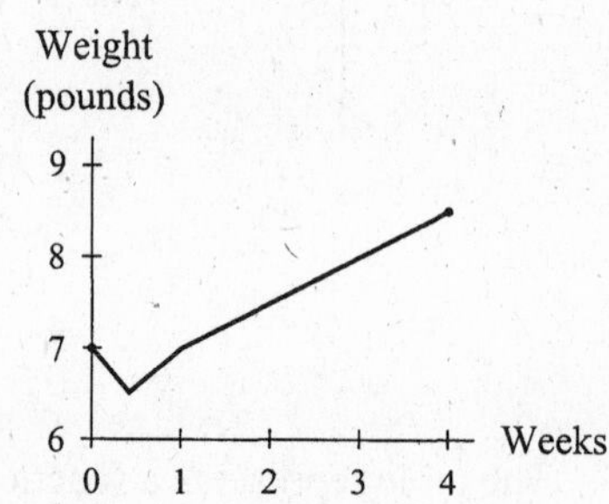

19. **a.** From the graph, the value is approximately \$9000.

b. From the graph, the monthly payment is approximately \$340.

c. From the graph, the payment for a \$15,000 car is approximately \$320, and the payment for a \$20,000 car is approximately \$425. The amount of increase is approximately \$425 – \$320 = \$105.

d. The graph would pass through (0, 0) but would lie below the 10% interest rate graph because the same monthly payment would pay for a smaller loan amount.

21. **a.** From the graph, it was approximately 2.6%

b. Cost-of-living increase was greatest in 2001 at 3.5%.

c. In 2004.

d. Benefits increased, but they increased by a lower percentage in 2003 than 2002. **Note: The answer in the text has incorrect years.**

23. $s = 5$: $t(5) = 3(5) + 6 = 21$
$s = 10$: $t(10) = 3(10) + 6 = 36$

25. $R(3) = 9.4(1.8^3) = 54.8208$

$R(0) = 9.4(1.8^0) = 9.4$

27. Solve $t(s) = 18$ using technology or algebraically as follows:

$$18 = 3s + 6$$
$$12 = 3s$$
$$s = 4$$

$t(s) = 18$ when $s = 4$.

Solve $t(s) = 0$ using technology or algebraically as follows:

$$0 = 3s + 6$$
$$-6 = 3s$$
$$s = -2$$

$t(s) = 0$ when $s = -2$.

29. Solve $R(w) = 9.4$ using technology or algebraically as follows:

$$9.4 = 9.4(1.8^w)$$
$$1 = (1.8^w)$$
$$\ln 1 = \ln(1.8^w)$$
$$0 = w\ln(1.8)$$
$$0 = w$$

$R(w) = 9.4$ when $w = 0$.

Solve $R(w) = 30$ using technology or algebraically as follows:

$$30 = 9.4(1.8^w)$$
$$\frac{30}{9.4} = (1.8^w)$$
$$\ln\left(\frac{30}{9.4}\right) = w\ln(1.8)$$
$$w = \frac{\ln\left(\frac{30}{9.4}\right)}{\ln(1.8)} \approx 1.974$$

$R(w) = 30$ when $w \approx 1.974$.

31. An input is given.

$$A(15) = 32e^{0.5(15)}$$
$$\approx 57{,}857.357$$

The corresponding output is approximately 57,857.357.

33. An output is given. Use technology to solve or find the solution algebraically as shown:

$$3.65 = 3(1.04^x)$$
$$\frac{3.65}{3} = (1.04^x)$$
$$\ln\left(\frac{3.65}{3}\right) = x\ln(1.04)$$
$$x = \frac{\ln\left(\frac{3.65}{3}\right)}{\ln(1.04)} \approx 5.000$$

The corresponding input is ≈ 5.000.

35. **a.** Profit = Revenue – Cost = \$5.3 million – \$4.2 million = \$1.1 million

b. $P(x) = R(x) - T(x)$ million dollars gives the profit from the production and sale of x units.

37. **a.** Manipulating the expression Profit = Revenue – Cost gives us Cost = Revenue – Profit.

Therefore, Cost = \$35 million - \$19 million = \$16 million

b. $T(x) = R(x) - P(x)$ million dollars gives the total cost for the company during the xth quarter.

39. **a.** Average Cost = $\frac{\$19.50}{150 \text{ bottles}} = \$0.13$ per bottle

b. $A(x) = \frac{C(x)}{x}$ dollars per unit gives the average cost for producing x units.

41. **a.** $r(y) = 100 \cdot \frac{P(y)}{D(y)}$

b. percent

43. **a.** Because the $S(t)$ is measured in dollars and $C(t)$ is measured in thousands of dollars, we must convert one of them to match the other. We choose to multiply $C(t)$ by 1000 in order to convert it to dollars:

$$\begin{aligned} Y(t) &= S(t) + 1000C(t) + 650{,}000 \\ &= 69{,}375t + 380{,}208 + 1000(-31.67t^2 + 137.15t + 233.5) + 650{,}000 \\ &= -31{,}670t^2 + 206{,}525t + 1{,}263{,}708 \text{ dollars} \end{aligned}$$

gives the VP's total yearly salary package t years after 1996, $0 \le t \le 2$.

b. $T(1) = -31{,}670(1)^2 + 206{,}525(1) + 1{,}263{,}708 = \$1{,}438{,}563$

45. **a.** The average credit card debt per cardholder can be calculated as

$$\text{average debt per card holder} = \frac{\text{total debt}}{\text{number of cardholders}}.$$

So average debt can be expressed by the function

$$\begin{aligned} A(y) &= \frac{42.4y + 219.5 \text{ billion dollars}}{1.7y + 140.3 \text{ million cardholders}} \\ &= \frac{42.4y + 219.5}{1.7y + 140.3} \text{ thousand dollars per cardholder} \end{aligned}$$

gives the average credit card debt y years after 1990, $8 \le y \le 15$.

b. In 2005, the average debt of a cardholder is $A(15) \approx \$5.16$ thousand per cardholder.

47. $c(x) = n(x)p(x) = (-0.034x^3 + 1.331x^2 + 9.913x + 164.447)(-0.183x^2 + 2.891x + 20.215)$ cesarean-section deliveries performed during the 1980s x years after 1980 on women who were 35 years of age or older.

49. The functions cannot be combined by function composition, because neither function has output that can be used as an input to the other function.

51. The functions can be combined because outputs from R can be used as inputs to D.
$(D \circ R)(x) = D(R(x))$ = revenue in dollars from the sale of x soccer uniforms

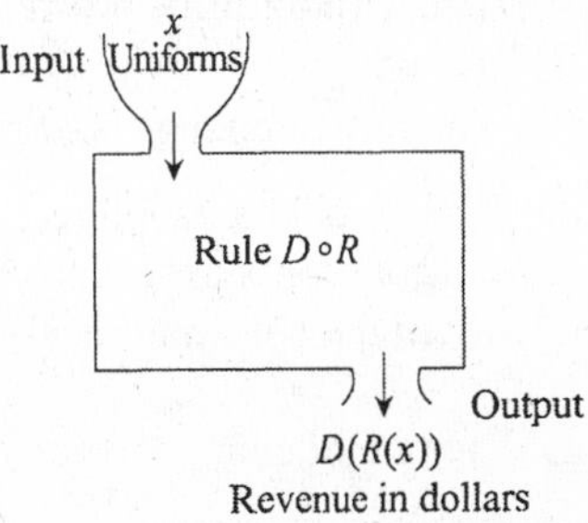

53. $h(p(t)) = h(1 + 3e^{-0.5t}) = \dfrac{4}{1 + 3e^{-0.5t}}$

55. $c(x(t)) = c(4 - 6t) = 3(4 - 6t)^2 - 2(4 - 6t) + 5$

57. *One possible answer:* For us to combine functions using addition, multiplication, or division, the input units must be the same. For us to combine functions using addition, the output units must be the same or must be able to be converted to the same unit (i.e., dollars and thousand dollars). For us to combine functions using multiplication or division, the output units must be such that when combined, they have a practical interpretation. For us to combine functions using composition the output units of one function must match the input units of the other function.

Section 1.2 Linear Functions and Models

1. a. 3 dollars per year
b. $f(0) = 3(0)+5 = 5$ dollars

3. a. 2 thousand dollars per hundred units or 20 dollars per unit
b. $r(0) = 2(0) - 4.5 = -4.5$ thousand dollars

5. a. the slope is negative (-2)
b. $f(x)$ is deceasing
c. -4

7. a. the slope is negative (-3)
b. $k(r)$ is deceasing
c. 7

9. $C(x) = 0.30x + 1.50$ dollars is the total cost for x units.

11. $S(h) = 0.25h + 3$ inches of snow on the ground h hours since noon.

13. a. The profit function is decreasing. The slope of the graph is negative.
b. $\text{Slope} = \dfrac{0 - 2.5 \text{ million \$}}{5 - 0 \text{ years}}$
$= -\$0.5$ million per year
The corporation's profit was declining by a half a million dollars per year during the 5-year period.
c. The rate of change is $-\$0.5$ million per year.
d. The vertical axis intercept is approximately \$2.5 million. This is the value of the corporation's profit in year zero. The horizontal axis intercept is 5 years. This is the time when the corporation's profit is zero.

15. a. The rate of change is positive and *D(t)* is increasing.
b. 382.5 donors per year

c.

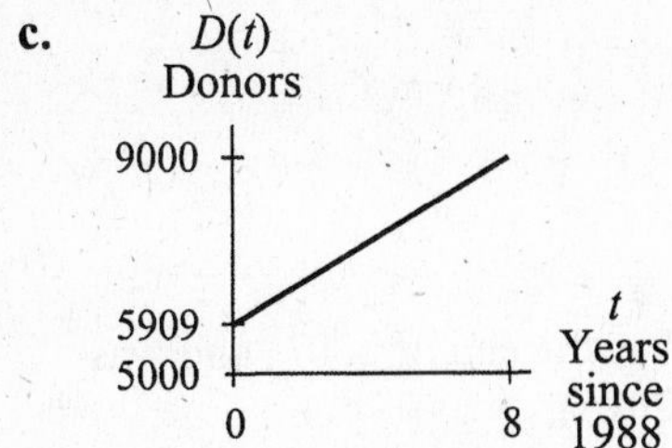

The slope of the graph is 382.5 donors per year.

d. $D(0) = 5909$. In 1988, there were 5,909 donors.

17. a. Rate of change of revenue $= \dfrac{\$2484.8\text{–}\$2128.1 \text{ million}}{2004\text{–}2003} = \$356.7$ million per year

b. Because $\dfrac{356.7}{4} = 89.175$, the revenue increased by \$89.175 million during each quarter of 1998.

c. Add \$356.7 million to each revenue amount to find the next year's revenue.

Year	Revenue (millions of dollars)
2003	2128.1
2004	2484.8
2005	2841.5
2006	3198.2

d. $R(y) = 356.7y + 2128.1$ million dollars gives the revenue y years after 2003, $0 \le y \le 3$.

19. a.

$$\begin{aligned}\text{Rate of change} &= \frac{\$112{,}000 - \$97{,}500}{2007 - 2000} \\ &= \frac{\$14{,}500}{7 \text{ years}} \\ &\approx \$2071.43 \text{ per year}\end{aligned}$$

That is approximately \$2071 per year.

b. $\$112{,}000 + (3 \text{ years})(\$2071 \text{ per year})) \approx \$118{,}214$

c. Let t be the number of years after the end of 2000. Then the predicted value is given by $V(t) = 2071t + 97{,}500$ dollars.

We begin the investigation by solving the following equation:

$$\begin{aligned}V(t) &= 100{,}000 \\ 2071.429t + 97{,}500 &= 100{,}000 \\ 2071.429t &= 2500 \\ t &\approx 1.207\end{aligned}$$

We continue by solving the following equation:

$$V(t)=150{,}000$$
$$2071.429t+97{,}500=150{,}000$$
$$2071.429t=52{,}500$$
$$t\approx 25.345$$

The value was \$100,000 in early 2002 (assuming $t = 1$ is the end of 2001) and was \$150,000 in mid-2026 (assuming $t = 25$ is the end of 2025).

d. $V(t) = 2071.429t + 95{,}000$ dollars gives the value of the house t years after the end of 2000, $0 \le t \le 7$.
2005: $V(5) = 2071.429(5) + 95{,}000 \approx \$105{,}357$
The model assumes the rate of increase of the market value remains constant. This assumption is not necessarily true. (In some markets, home prices fluctuate wildly.)

21. a.
$$\text{Rate of increase} = \frac{\$2040\text{–}\$1719 \text{ billion}}{2003\text{–}2000}$$
$$= \frac{\$321 \text{ billion}}{3 \text{ years}} = \$107 \text{ billion / year}$$

b. \$2040 billion + (3 years)(\$107 billion per year) = \$2361 billion

c. *One possible answer:* No, the rate at which consumers are willing to borrow money will fluctuate with changes in the sate of the national economy.

d. If t is the number of years after 2000, then $C(t) = 107t + 1719$ billion dollars. Because \$3 trillion = \$3000 billion, solve $C(t) = 3000$.
$$107t+1719=3000$$
$$107t=1281$$
$$t\approx 11.972$$

The linear model predicts that consumer credit will reach \$3 trillion approximately 12 years after 2000, in the year 2012.

23. a. 78 million people per year

b. $P(t) = 6 + 0.078t$ billion people gives the world's population t years after the beginning of 2000.

c. Setting the model in part *b* equal to 12 and solving for t yields $t \approx 76.9$ years after the beginning of 2000, which corresponds to near the end of 2076. The article estimates the world population will be 12 billion in 2050.

d. The prediction in part *c* assumes that the world will grow at a constant rate of 78 million people per year between now and 2076.

25. Note: Answers in text for this activity are incorrect.

a. The points in the scatter plot appear to lie in a line.

b. The first differences (or changes in output) are \$0.63 – \$0.39 = \$0.24, \$0.87 – \$0.63 = \$0.24, and so on. The first differences, listed in order are {0.24, 0.24, 0.24, 0.28, 0.2, 0.24, 0.24, 0.24}.

c. Using technology yields the model: $P(w) = 0.24w + 0.154$ dollars is the first-class domestic postage rate for weights not exceeding w ounces, $1 \le w \le 9$.

27. a. *Two possible models are:*

$S1(x) = 499.3x - 976{,}088.3$ students is the enrollment in year x, $1965 \le x \le 1969$.

$S2(x) = 499.3x + 5036.2$ students is the enrollment x years after 1965, $0 \le x \le 4$.

b. The enrollment in 1970 can be estimated as

$S1(1970) \approx 7533$ students

$S2(5) \approx 7533$ students

c. Because $7533 - 8038 = -505$, the estimate is 505 students lower than the actual enrollment. Answers vary on whether the error is significant. The error represents approximately 6% of the actual enrollment. For a school the size of the one in this activity, an error of 500 students could mean a significant increase in student housing and faculty loads.

d. These models should not be used to predict enrollment in the year 2000, because the data are too far removed from 2000 to be of any value in such a prediction.

29. a. 1990: $e(0) = 0.107(0) + 5.11 = 5.11$ million gigagrams
1997: $e(7) = 0.107(7) + 5.11 = 5.859$ million gigagrams
2022: $e(12) = 0.107(12) + 5.11 = 6.394$ million gigagrams

b. 0.107 million gigagrams per year

c. $e(22) = 0.107(22) + 5.11 = 7.464$ million gigagrams

31. *One possible answer:* The equation makes it possible for us to use mathematics to answer numerical questions concerning the situation being modeled. The units of measure on the output and the description (including units of measure) on the input makes it possible for us to interpret the numerical answers in the context of the situation. The interval of inputs helps us to know when we are extrapolating.

Section 1.3 Exponential and Logarithmic Functions and Models

1. $f(x) = 2(1.3^x)$ is the black graph. $f(x) = 2(0.7^x)$ is the teal graph.

3. $f(x) = 3(1.2^x)$ is the teal graph. $f(x) = 2(1.4^x)$ is the black graph.

5. $f(x) = 2\ln x$ is the teal graph. $f(x) = -2\ln x$ is the black graph.

7. $f(x) = 2\ln x$ is the teal graph. $f(x) = 4\ln x$ is the black graph.

9. Because $1.05 = 1 + 0.05$, f is increasing with a 5% change in output for every unit of input.

11. Because $0.87 = 1 - 0.13$, y is decreasing with a 13% change in output for every unit of input.

13. Because $0.61 = 1 - 0.39$, the number of bacteria declines by 39% each hour.

15. a. With starting value $a = 4.81$ quadrillion Btu and the parameter b calculated as $b = 1 + 0.0547$, the model is $P(t) = 4.81(1.0547^t)$ quadrillion Btu is the projected amount of petroleum product imports t years after 2005, $0 \le t \le 15$.

b. Solving the equation $10 = 4.81(1.0547^t)$ for t yields $t \approx 13.7$ years after 2005. Thus petroleum product imports will exceed 10 quadrillion Btu for the first time in 2019 (assuming $t = 13$ is the end of 2018).

c. $P(t)$ is an increasing exponential function so the projected petroleum product imports increase without bound as time increases.

17. a. Using starting value $a = 3.3$ and calculating b as $b = 1 - 0.0146 = 0.9854$ produces the model $W(t) = 3.3(0.9854^t)$ workers per Social Security beneficiary t years after 1996 between 1996 and 2030.

b. $W(34) \approx 2.00$ workers per beneficiary. Fewer workers per beneficiary will mean that Social Security will have to find other means of supplementing payments rather than relying solely on Social Security withholdings from workers' wages.

19. a. Sales declined by 520,000 – 210,000 = 310,000 tapes, which is a $\frac{310{,}000}{520{,}000} \cdot 100\% \approx 59.6\%$ decrease. The monthly decline was 59.6%.

b. Because 1 – 0.596 = 0.404, the model is $S(t) = 520{,}000(0.404^t)$ videotapes sold per month t months after publicity was discontinued.

c. 3 months:

$$S(3) = 520{,}000\left(0.404^3\right)$$
$$\approx 34{,}288 \text{ tapes per month}$$

12 months:

$$S(12) = 520{,}000\left(0.404^{12}\right)$$
$$\approx 10 \text{ tapes per month}$$

21. a. $T(y) = 0.002(1.397^y)$ million transistors in Intel processor chips y years after 1970, $1 \le y \le 35$.

b. $(1.39746 - 1)100 = 39.746\%$ each year

c. Yes, the data seem to support Moore's Law. If anything the number of transistors has been doubling faster than every 2 years.

23. a. $L(x) = 0.845x + 0.790$ gallons per person per year gives the per capita consumption of bottled water in the United States x years after 1980, $0 \le x \le 23$.

$E(x) = 2.714(1.099^x)$ gallons per person per year gives the per capita consumption of bottled water in the United States x years after 1980, $0 \le x \le 23$.

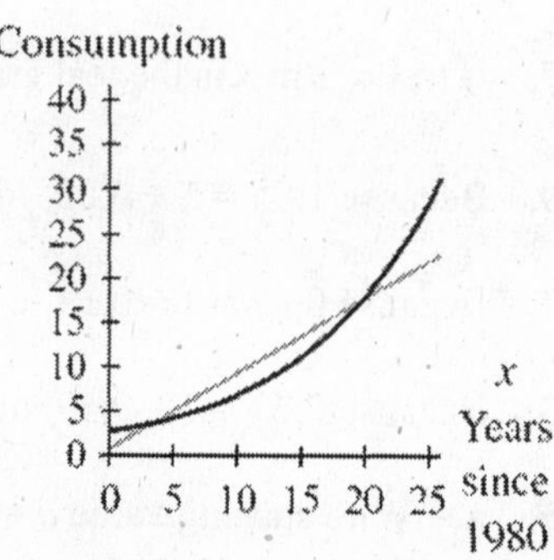

The exponential model appears to better describe the per capita bottled water consumption on the given interval.

b. The linear model describes a rate of increase of approximately 0.845 gallons per person per year. The exponential model describes a $(1.099 - 1)(100\%) \approx 9.927\%$ per year percentage increase.

c. Linear: $L(25) = 0.845(25) + 0.790 \approx 21.9$ gallons per person per year

Exponential: $E(25) = 2.714(1.099^{25}) \approx 28.7$ gallons per person per year

d. Solving $L(x) = 25$ yields $x \approx 28.651$

According to the linear model, water consumption exceeded 25 gallons per person in 2009.

Solving $E(x) = 25$ yields $x \approx 23.521$
According to the exponential model, water consumption exceeded 25 gallons per person in 2004.

25. a. The model for the amount of radon gas after t hours when R_0 is the initial amount of radon is $R = R_0 b^t$. Using $R = 0.8R_0$ when $t = 30$, we have

$$0.8R_0 = R_0 b^{30}$$
$$0.8 = b^{30}$$
$$\ln 0.8 = \ln b^{30}$$
$$\ln 0.8 = 30 \ln b$$
$$\frac{\ln 0.8}{30} = \ln b$$
$$b = e^{\ln 0.8/30} \approx 0.992589$$

So we have, $R = R_0 e^{(\ln 0.8)t/30}$ Substituting $R = 0.5R_0$ gives

$$0.5R_0 = R_0 e^{(\ln 0.8)t/30}$$
$$0.5 = e^{(\ln 0.8)t/30}$$
$$\ln 0.5 = \frac{(\ln 0.8)t}{30}$$
$$t = \frac{30 \ln 0.5}{\ln 0.8} \approx 93.2 \text{ hours}$$

b. $R = R_0 e^{(\ln 0.8)t/30}$ after t hours. The output units are the same as those of the initial amount R_0.

c. $\lim_{t\to\infty} R_0 e^{(\ln 0.8)t/30} = R_0 \lim_{t\to\infty} e^{-0.007438t} = 0$
Given enough time, there will be no more radon present in the building.

27. We solve $\frac{1}{2}D_0 = D_0 e^{-0.0198t}$ for t to obtain a half-life of $t = 35$ hours.

29. **a.** $D(t) = 406.401(0.906^t)$ is the number of days that milk will keep when stored at a temperature of t degrees Fahrenheit, $30 \le t \le 70$.

b. $D(40) - D(37) \approx -2.7$; The milk will spoil approximately 2.7 days sooner.

c. $t(D) = 60.547 - 9.913 \ln D$ degrees Fahrenheit is the temperature at which milk should be stored in order for the milk to keep for D days, $0.5 \le D \le 24$.

d. $t(7) = 60.547 - 9.913 \ln(7) \approx 41.257$; The refrigerator should be set at 41.3 degrees Fahrenheit.

31. **a.** $R(y) = 8.435 - 0.639 \ \ln\ y$ percent gives the New Zealand bond rate for a maturity time of t years, $0.25 \le y \le 10$.

b. The model estimates 15-year bond rates as $R(15) \approx 6.70$,, which is 0.3 percentage point less than the fund manager's estimate.

c. $T(p) = 461{,}733.212(0.213^p)$ years is the time to maturity for a New Zealand bond with a p% rate, $7.10 \le p \le 9.40$.

d.

$$R(T(9.4)) \approx R(0.228169) \approx 9.4$$
$$R(T(7.5)) \approx R(4.294381) \approx 7.5$$
$$R(T(7.1)) \approx R(7.966118) \approx 7.1$$

$$T(R(2)) \approx T(7.992363) \approx 2$$
$$T(R(4)) \approx T(7.549248) \approx 4$$
$$T(R(10)) \approx T(6.963482) \approx 10$$

These calculations suggest that R and T are inverse functions because $R(T(p)) \approx p$ and $T(R(y)) \approx y$.

33. **a.** $C(d) = 1.182 + 2.216 \ln d$ μg/mL is the concentration of a drug in the bloodstream after d days, $1 \le d \le 17$.

b. $\lim\limits_{d \to \infty} C(d) \to \infty$ and $\lim\limits_{d \to 0^+} C(d) \to -\infty$

Note: Answer in text has incorrect limit notation.

c. *One possible answer:* The context tells us that the amount of concentration will continue to increase. The logarithmic model also indicates an increase and fits the end behavior suggested by the context.

d. $C(2) = 1.182 + 2.216 \ln 2$

≈ 2.7 μg/mL

35. **a.** $p(x) = -9.792 \cdot 10^{-5} - 0.434 \ln x$ is the pH of a solution, where x is the H_3O^+ concentration in moles per liter, $5.012 \cdot 10^{-9} \le x \le 3.981 \cdot 10^{-7}$.

b. $p(1.585 \cdot 10^{-3}) = -9.792 \cdot 10^{-5} - 0.434 \ln(1.585 \cdot 10^{-3}) \approx 2.8$

c. Solving the equation $5.0 = -9.792 \cdot 10^{-5} - 0.434 \ln x$ for x gives $x \approx 1.0 \cdot 10^{-5}$ moles per liter.

d. The pH of beer is approximately $p(3.162 \cdot 10^{-5}) \approx 4.5$, which means it is acidic.

37. *One possible answer:* For an exponential model the length of the input interval over which the output values either double or halve will be the same no matter where the interval starts. For a linear model the length of the input interval over which the output values double or halve is directly affected by the starting endpoint of the interval.

39. *One possible answer:* The end behavior of an exponential model in standard form ($f(x) = ab^x$) is that $f(x)$ increases or decreases without bound in one direction (as x approaches $\pm\infty$) and approaches zero in the other direction. By contrast, a logarithmic model increases without bound in one direction (as x approaches $\pm\infty$) and decreases without bound in the other direction.

Section 1.4 Logistic Functions and Models

1. The concave-up, decreasing shape could be either exponential or logarithmic.

3. The increasing, concave-down shape is that of a logarithmic function.

5. The scatter plot is none of these types. It exhibits change in concavity, so it cannot be linear, logarithmic, or exponential. It does not level off, so it is not logistic.

7. Note that $f(x) = \dfrac{L}{1 + Ae^{-Bx}}$ where
$L = 100$, $A = 9$, and $B = 0.78$. Because L, A, and B are positive, f is increasing to a limiting value of $L = 100$.

9. Note that $h(g) = \dfrac{L}{1 + Ae^{-Bg}}$ where
$L = 39.2$, $A = 0.8$, and $B = -3$. Because L and A are positive and B is negative, h is decreasing from a limiting value $L = 39.2$.

11. a. $C(t) = \dfrac{37.195}{1 + 21.374^{-0.183t}}$ countries in Europe, North America and South America t years after 1840, $0 \le s \le 40$. The model is a good fit.

b.

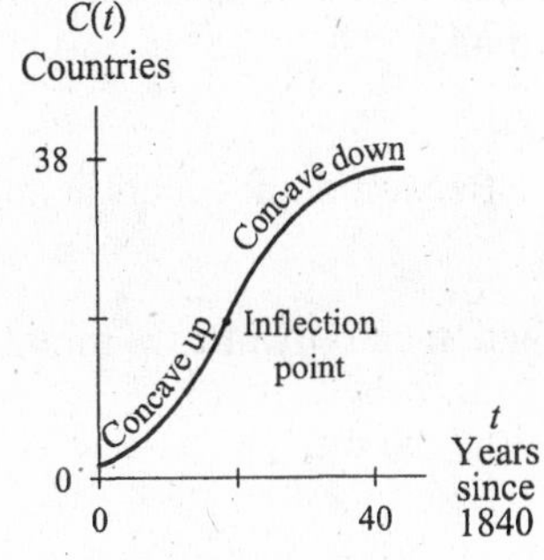

13. a,b. $N(t)=\dfrac{3015.991}{1+119.250e^{-1.024t}}$

Navy deaths t weeks after August 31, 1918, $0 \le t \le 13$.

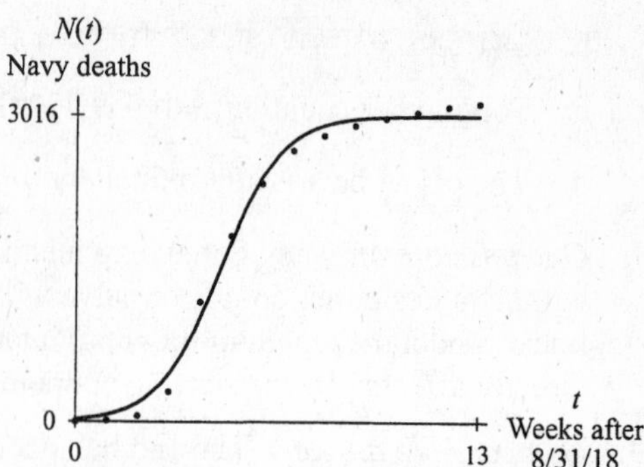

$A(t)=\dfrac{20{,}492.567}{1+518.860e^{-1.212t}}$

Army deaths t weeks after Sept. 7, 1918, $0 \le t \le 12$.

Answer in text has different alignment and input description. Either model is correct.

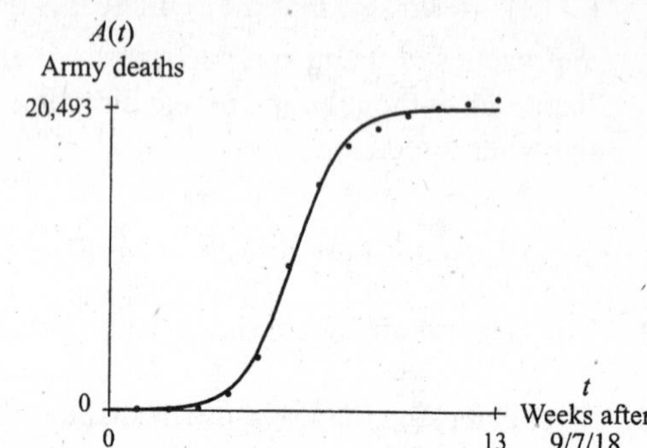

$C(t)=\dfrac{91{,}317.712}{1+176.272e^{-0.951129t}}$

Civilian deaths t weeks after Sept.14, 1918, $0 \le t \le 11$.

Answer in text has different alignment and input description. Either model is correct.

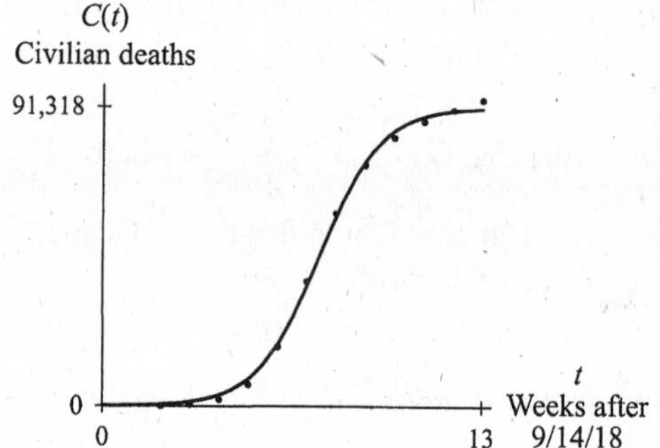

c. No. The models given in part *b* have limiting values less than the number of deaths in the table for November 30. The models are not good indicators of the ultimate number of deaths.

The data are concave down from January through April and concave up from April through June. This is not the concavity exhibited by a logistic model

15. a. The limiting value is approximately 2U/100 μg. The inflection point occurs at approximately 9 minutes. (Answers may vary.)

b. $A(m)=\dfrac{1.937}{1+29.064e^{-0.421110m}}$ U/100 μg gives the reaction activity after m minutes, $0 \le m \le 18$. The limiting value is approximately 1.94 U/100 μg.

c. $A(11)-A(7)\approx 1.51-0.77\approx 0.74$ U/100 μg

17. a. $P(x)=\dfrac{11.742}{1+154.546e^{-0.026x}}$ billion people gives the world's population x years after 1800, $4 \le x \le 271$. The equation is a good fit for the later (1960–2071) data but a poor fit for the early (1800–1960) data.

b. According to the model, the world population will level off at 11.7 billion. This is probably not an accurate prediction of future world population.

c. 1850: $P(50) \approx 0.266$ billion people
1990: $P(190) \approx 5.320$ billion people
The model does a poor job of estimating the 1850 population and a good job of estimating the 1990 population.

19. a. $g(t)$ is concave up.

b. $\lim\limits_{t\to-\infty} g(t)=0$; $\lim\limits_{t\to\infty} g(t)=\infty$

c. As t decreases without bound, g approaches zero. As t increases without bound, g also increases without bound.

21. a. $y(x)$ is concave down

b. $\lim\limits_{x\to0^+} y(x)=-\infty$; $\lim\limits_{x\to\infty} y(x)=\infty$

c. As x approaches zero from the right-hand side, y decreases without bound. As x increases without bound, y also increases without bound.

23. a. $l(t)$ is concave up from $-\infty$ to approximately $x = 0$ and then is concave down.

b. $\lim\limits_{t\to-\infty} l(t)=0$; $\lim\limits_{t\to\infty} l(t)=52$

c. As t decreases without bound, l approaches zero. As t increases without bound, l approaches the limiting value of 52.

25. a. $n(k)$ is concave up

b. $\lim\limits_{k\to\pm\infty} n(k)=\infty$

c. As k increases or decreases without bound, $n(k)$ increases without bound.

27. a. $C(q)$ is concave up from $-\infty$ to approximately $q \approx 0.8$ and then is concave down.

b. $\lim\limits_{q\to-\infty} C(q)=\infty$; $\lim\limits_{q\to\infty} C(q)=-\infty$

c. As q decreases without bound, $C(q)$ increases without bound. As q increases without bound, $C(q)$ decreases without bound.

29. *One possible answer:* A logistic equation of the form $f(x)=\dfrac{L}{1+Ae^{-Bx}}$ is unlike either the exponential or logarithmic equations in that it is bounded above and below so that when $B>0$, $\lim\limits_{x\to-\infty} f(x)=0$ and $\lim\limits_{x\to\infty} f(x)=L$, or when $B<0$, $\lim\limits_{x\to-\infty} f(x)=L$ and $\lim\limits_{x\to\infty} f(x)=0$. An exponential equation is bounded only in one direction and is unbounded in the other. A logarithmic equation must have positive input and increases or decreases unbounded as its input increases without bound.

Section 1.5 Polynomial Functions and Models

1. Concave up, decreasing from $x = 0.75$ to $x = 3$, increasing from $x = 3$ to $x = 4$

3. Concave up, decreasing from $x = 13.5$ to $x = 18$, increasing $x = 18$ to $x = 22.5$

5. Concave down, always decreasing

7.

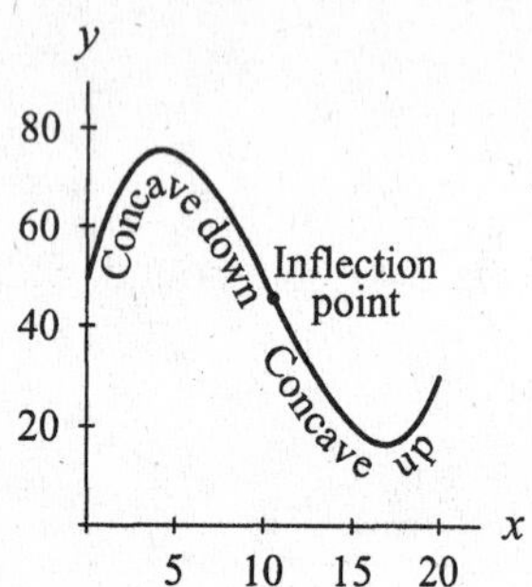

9.

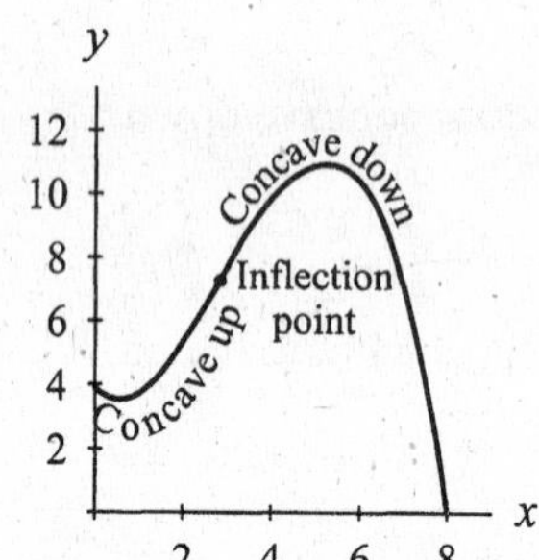

11.

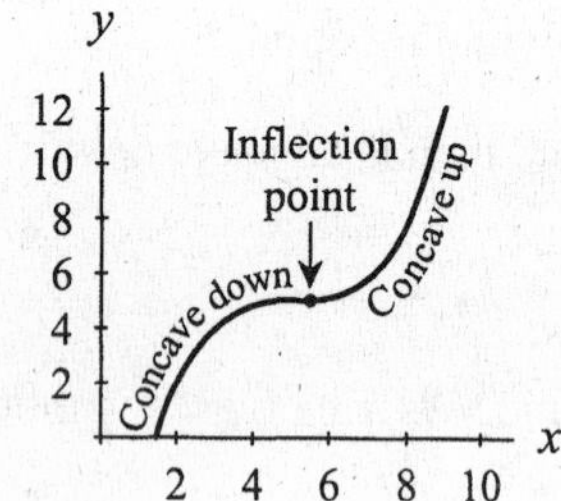

13. a. Calculate the first and second differences for the height data:

128 140 144 140 128 108 80

12 4 – 4 – 12 – 20 – 28

– 8 – 8 – 8 – 8 – 8

Second differences are constant, so the data are quadratic.

b. Work from bottom to top to continue the pattern above.

108 80 44 0

– 28 – 36 – 44

– 8 – 8

After 3.5 seconds the height is 44 feet. After 4 seconds the height is 0 feet.

c. $H(s) = -16s^2 + 32s + 128$ feet is the height of the missile after s seconds, $0 \le s \le 4$.

d. Solving $-16s^2 + 32s + 128 = 0$ yields $s = -2$ and $s = 4$. Because negative times values do not make sense in this context, we conclude that the missile hits the water after 4 seconds.

15. a. The first differences in the ages are $20.8 - 20.3 = 0.5$, $22 - 20.8 = 1.2$, and $23.9 - 22 = 1.9$, so the second differences are $1.2 - 0.5 = 0.7$ and $1.9 - 1.2 = 0.7$. Because the input data are evenly spaced and the second differences are constant, the data are perfectly quadratic.

b. The next first difference should be $1.9 + 0.7 = 2.6$, so the median age for 2000 should be $23.9 + 2.6 = 26.5$ years.

c. $A(x) = 0.0035x^2 - 0.405x + 32$ years is the median age at first marriage of females in the United States x years after 1900, $60 \le x \le 90$.

d. $A(100) = 0.0035(100^2) - 0.405(100) + 32 = 26.5$ years of age. Yes, this answer is the same as the answer to part *b*.

17. a. $D(x) = 0.025x^2 - 2.021x + 43.78$ deaths per thousand people is the death rate in 1998 for the United States for an age of x years, $40 \le x \le 65$. The equation appears to be a good fit.

b.

Age	Model prediction	Actual rate
51	$D(51) \approx 4.6$	4.7
52	$D(52) \approx 5.2$	5.1
53	$D(53) \approx 5.7$	5.6
57	$D(57) \approx 8.4$	8.1
59	$D(59) \approx 10.1$	9.7
63	$D(63) \approx 14.0$	14.1
70	$D(70) \approx 22.7$	25.5
75	$D(75) \approx 30.5$	38.0
80	$D(80) \approx 39.4$	59.2

c. The model is more accurate when used for interpolation than when used for extrapolation.

19. 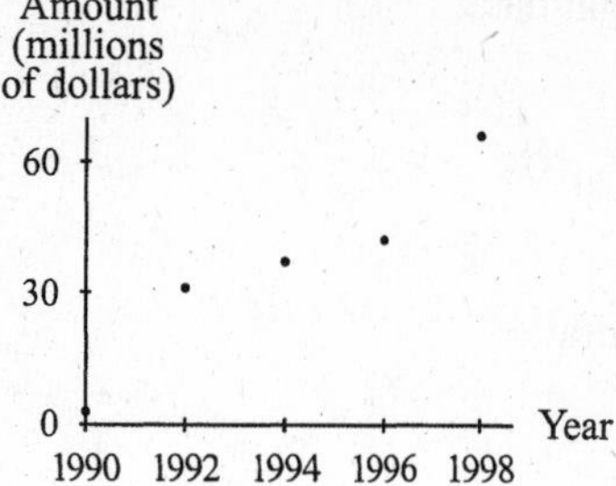

The scatter plot indicates an inflection point and does not indicate a limiting value.

b. $A(t) = 0.427t^3 - 5.286t^2 + 22.827t + 3.014$ million dollars spent t years after 1990, $0 \le t \le 8$.

c. 1993: $A(3) \approx \$35$ million
1999: $A(9) \approx \$92$ million
The 1993 estimate is more likely to be accurate because it is an interpolation rather than an extrapolation.

d. The 1993 estimate exceeded the actual amount by \$1 million. The 1999 estimate is \$7 million short of the actual amount. These figures confirm the statements in part *c*.

21. a. The scatter plot suggests an inflection point, a relative maximum, and a relative minimum.

b. $G(x) = (5.051 \cdot 10^{-5})x^3 - 0.007x^2 + 0.085x + 105.027$ males per 100 females is the gender ration in the United States x years after 1900, $0 \le x \le 100$. The graph of G rises beyond 2000. Answers will vary, depending on whether the gender ratio will rise as indicated by the model.

23. a. The number of females and males is approximately equal for 40-year-olds.

b. Using "under age" as 0 and "100 and over" as 100 for modeling, but not prediction purposes, we get the following:
Cubic model: $C(a) = (-19.590 \cdot 10^{-5})a^3 + (18.421 \cdot 10^{-4})a^2 + 0.037a + 104.601$ males per 100 females gives the gender ration in the United States for individuals who are a years old, $0 \le a \le 100$.

Logistic model: $L(a) = \dfrac{104.3}{1 + (9.817 \cdot 10^{-4})e^{0.082a}}$ males per 100 females gives the gender ration in the United States for individuals who are a years old, $0 \le a \le 100$.

The logistic equation fits the data better than the cubic equation, especially for ages above 60.

c. Solving $L(a) = 50$ gives $a \approx 85.6$ years of age. Among 86-year-olds there are approximately twice as many women as men. This implies that men die younger than women.

25. *One possible answer:* A graph of $y = ax^2 + bx + c$ will be concave up when a is positive. It will be decreasing to a minimum, after which it will be increasing. When a is negative, a graph of $y = ax^2 + bx + c$ will be concave down—increasing to a maximum and then decreasing.

A graph of $y = ax^3 + bx^2 + cx + d$ could take on one of four forms. If a is positive, a graph could be concave down, increasing to an inflection point and then concave up, increasing; or it could be concave down and increasing to a maximum and then decreasing to an inflection point after which it would be concave up and decreasing to a minimum and then increasing. On the other hand, if a is negative, a graph could be concave up, decreasing to an inflection point and then concave down, decreasing; or it could be concave up and decreasing to a minimum and then increasing to an inflection point after which it would be concave down and increasing to a maximum and then decreasing.

Chapter 1 Concept Review

1. **a.** The scatter plot is concave up.

b. The scatter plot appears to be increasing without bound as x approaches $\pm\infty$.

c. Quadratic

d. $\lim_{x \to \pm\infty} f(x) = \infty$

2. **a.** The scatter plot is increasing, concave up.

b. End behavior to the left is not apparent from the scatter plot. As x increases, the scatter plot appears to be increasing without bound.

c. Either quadratic or exponential (shifted up 8)

d. Quadratic: $\lim_{x \to \pm\infty} f(x) = \infty$

Exponential: $\lim_{x \to -\infty} f(x) = 8$, and $\lim_{x \to \infty} f(x) = \infty$

3. **a.** The scatter plot does not indicate any curvature.

b. The scatter plot appears to be increasing without bound as x approaehes $-\infty$ and decreasing without bound as x approaches ∞.

c. Linear

d. $\lim_{x \to -\infty} f(x) = \infty$, and $\lim_{x \to \infty} f(x) = -\infty$

4. **a.** The scatter plot is increasing from 0 to 2 and from 4 to 6. It in concave down over 0 to 3 and concave up from 3 to 6. It appears to have an inflection point near $x = 3$.

b. The scatter plot appears to be decreasing without bound as x approaches $-\infty$ and increasing without bound as x approaches ∞.

c. Cubic

d. $\lim_{x \to -\infty} f(x) = -\infty$, and $\lim_{x \to \infty} f(x) = \infty$

5. **a.** The scatter plot is concave up from $x = 0$ to $x = 3$ and concave down from $x = 3$ to $x = 6$.
   b. The scatter plot appears to be increasing toward 40 as x approaches ∞ and decreasing toward zero as x approaches $-\infty$.
   c. Logistic
   d. $\lim_{x\to-\infty} f(x) = 0$, and $\lim_{x\to\infty} f(x) = 40$

6. **a.** The scatter plot is increasing, concave up.
   b. The scatter plot appears to be decreasing without bound as x approaches 0 from the right and increasing without bound but more and more slowly as x approaches ∞.
   c. Logarithmic
   d. $\lim_{x\to0} f(x) = -\infty$, and $\lim_{x\to\infty} f(x) = \infty$

7. **a.**

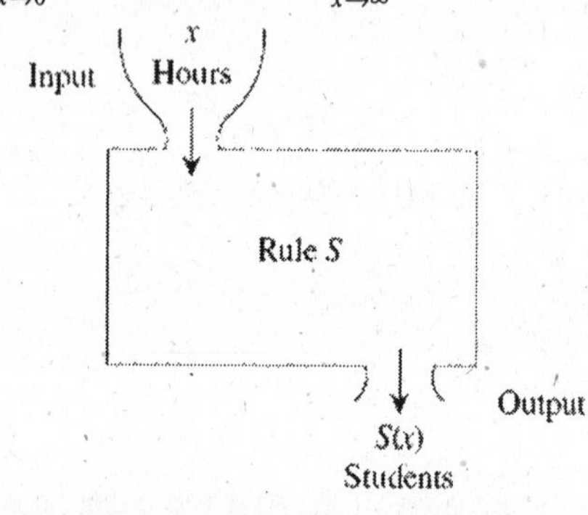

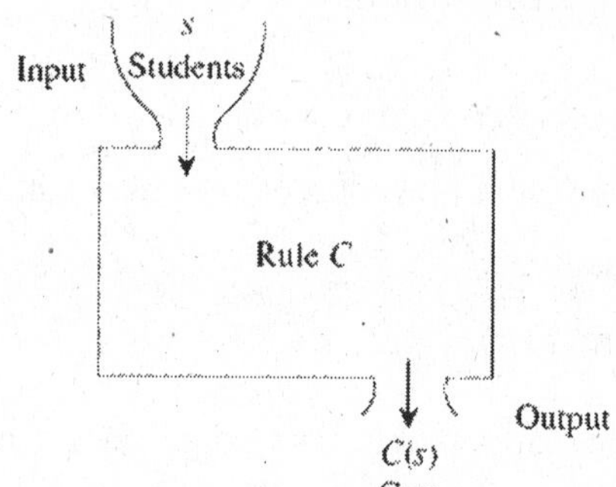

   b. Composition
   c. $(C \circ S)(x) = C(S(x))$
   d. Input: hours; Output: cars

8. **a.**

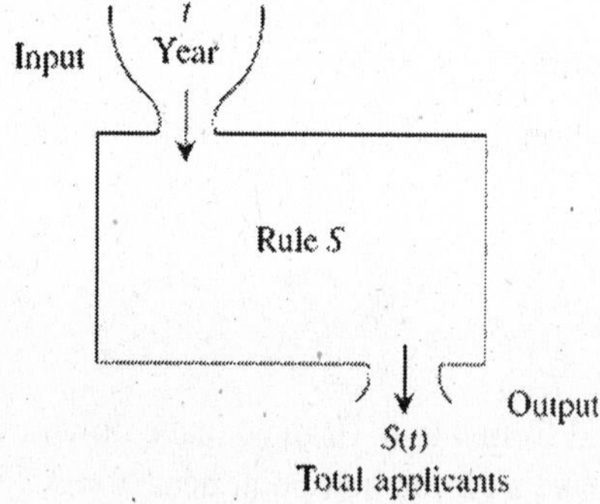

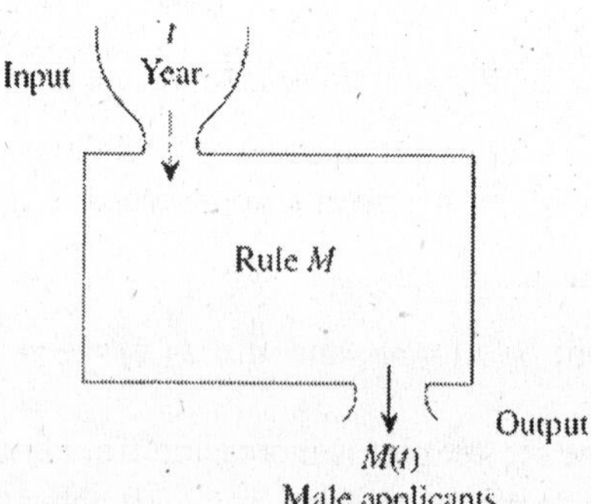

   b. Subtraction
   c. $F(t) = S(t) - M(t)$
   d. Input: none; Output: applicants

9. a.

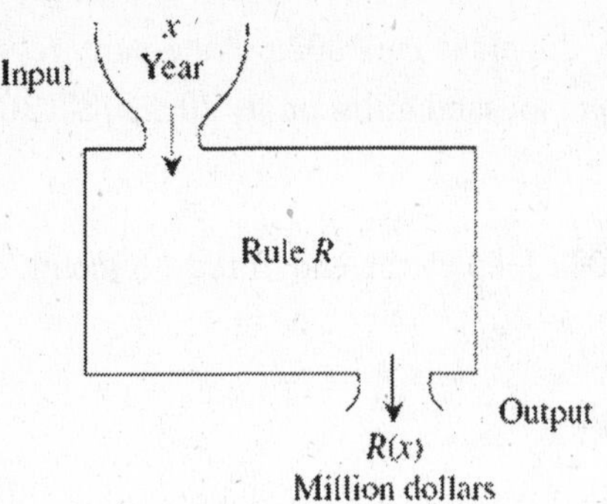

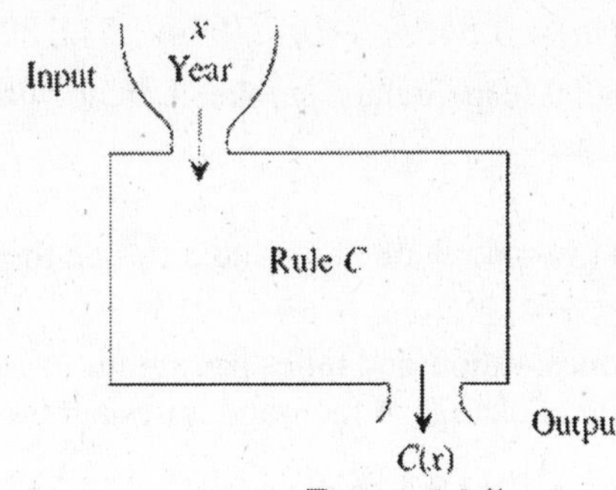

b. The profit function can be constructed from the revenue and cost functions.

c. $P(x) = 1,000,000R(x) - 1000C(x)$ (Answers may vary.)

d. $P(x) = 1,000,000R(x) - 1000C(x)$ **dollars is the** company's profit after it has been in business x years.

10. a. Multiplication

b. $P(p) = \dfrac{40p}{1 + 0.03e^{0.4p}}$

c. $P(p) = \dfrac{40p}{1 + 0.03e^{0.4p}}$ million dollars is the revenue when p dollars is the price of a pound of beef.

11. a. $C(x) = 2.2(1.021^x)$ million children living with their grandparents x years after 1970, $0 \le x \le 27$.

b. Approximately 2.1% each year

c. $C(x) = 5$ when $x \approx 38.717$. According to the model, the number of children living with their grandparents will reach 5 million in 2009.

d. $C(x) = 4.4$ when $x \approx 32.689$ years; The number of children living with their grandparents in 1970 will have doubled by 2003.

12. a. decreasing; The number 0.88 is less than 1 which indicates a decreasing exponential function.

b. \$7: $D(7) \approx 2.55$ trillion pounds
\$14: $D(14) \approx 1.04$ trillion pounds
\$21: $D(21) \approx 0.43$ trillion pounds

c. Logarithmic

d. $P(d) = 14.336 - 7.823 \ln d$ dollars per pound is the market price when d trillion pounds of fish is demanded.

13. a. The scatter plot indicates a single concavity, which indicates that a quadratic or exponential model could be used. In this instance, an exponential model will not fit the original data set well because the output data are not approaching zero.

$J(m) = 0.546m^2 - 141.763m + 21{,}382.5$ dollars is the 2002 private-party resale value of a 2000 Jeep Grand Cherokee Laredo with m thousand miles on it, $20 \le m \le 120$.

b. $J(52) \approx \$15{,}487$

c. $M(x) = 4x + 68$ thousand miles on the 2002 Jeep by the end of the xth month of 2002, $0 \le x \le 12$.
Slope: 4 thousand miles per month
Rate of change: 4 thousand miles per month

d. $J(M(x)) = 0.546(4x+68)^2 - 141.763(4x+68) + 21{,}382.5$ dollars is the 2002 private-party resale value of a 2000 Jeep Grand Cherokee Laredo at the end of the xth month of 2002, $0 \le x \le 12$.

14. a. The scatter plot is concave up to the left of 8 and concave down to the right of 8. There is an inflection point near (8, 21,200).
Logistic

b. As x decreases the data approach 0. As x increases the data approach a limiting value.

c. Logistic

d. $P(m) = \dfrac{42{,}183.911}{1 + 21{,}484.252e^{-1.249m}}$ polio cases gives the cumulative number of polio cases diagnosed in the United States by the mth month of 1949.

Chapter 2
Describing Change: Rates

Section 2.1 Change, Percentage Change, and Average Rates of Change

1. $\dfrac{\$2.30}{5 \text{ days}} = \$0.46$ per day

The stock price rose an average of 46 cents per day during the 5-day period.

3. $\dfrac{\$25{,}000}{3 \text{ months}} \approx \$8333.33$ per month

The company lost an average of $8333.33 per month during the past three months.

5. Change: 56.5 million − 55.4 million = 1.1 million

From 2004 to 2005, the number of passengers flown by Northwest Airlines increased by 1.1 million.

Percentage change: $\left(\dfrac{56.5 \text{ million} - 55.4 \text{ million}}{55.4 \text{ million}}\right)100\% \approx 1.986\%$

From 2004 to 2005, the number of passengers flown by Northwest Airlines increased by approximately 2%.

Average rate of change:

$\dfrac{56.5 \text{ million} - 55.4 \text{ million}}{2005 - 2004} = \dfrac{1.1 \text{ million}}{1 \text{ year}} = 1.1$ million passengers per year

Between 2004 and 2005, the number of passengers flown by Northwest Airlines increased an average of 1.1 million passengers per year.

7. Change: 2434 − 362 = 2072 thousand people

Between 1930 and 2000, the American Indian, Eskimo, and Aleut population in the U.S. increased by 2,072,000 people.

Percentage change: $\left(\dfrac{2072 \text{ thousand people}}{362 \text{ thousand people}}\right)100\% \approx 672\%$

Between 1930 and 2000, the American Indian, Eskimo, and Aleut population in the U.S. increased by approximately 672%. In other words, the population in 2000 was approximately six and a half times what it was in 1930.

Average rate of change:

$\dfrac{2072 \text{ thousand people}}{2000 - 1930} = \dfrac{2434 \text{ thousand people}}{70 \text{ years}} = 29.6$ thousand people per year

Between 1930 and 2000, the American Indian, Eskimo and Aleut population in the U.S. increased, on average, by 29,600 people per year.

9. a. From the graph we see that October 1987 had 22 trading days. (We must remember that the first day gives us the initial value for the interval but cannot be numbered in the days over which our average rate of change is calculated.)

$$\text{Average rate of change} = \frac{303.4 \text{ million} - 193.2 \text{ million}}{22 - 1} = \frac{110.2 \text{ million shares}}{21 \text{ trading days}}$$

$$\approx 5.3 \text{ million shares per trading day}$$

Note: the answer given in the text is calculated using the number of days between October 1 and October 30 (29 calendar days instead of 21 trading days) giving

$$\frac{\textbf{110.2 million shares}}{\textbf{29 days}} \approx \textbf{3.8 million shares per day}.$$

During October 1987, the number of shares traded each day was increased at an average rate of approximately 5 million shares per trading day.

$$\text{Percentage change} = \frac{303.4 - 193.2 \text{ million shares}}{193.2 \text{ million shares}} \cdot 100\% \approx 57\%$$

The number of shares traded on the last trading day of October 1987 was 57% more than the number traded on the first trading day of that month.

b.

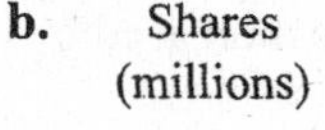

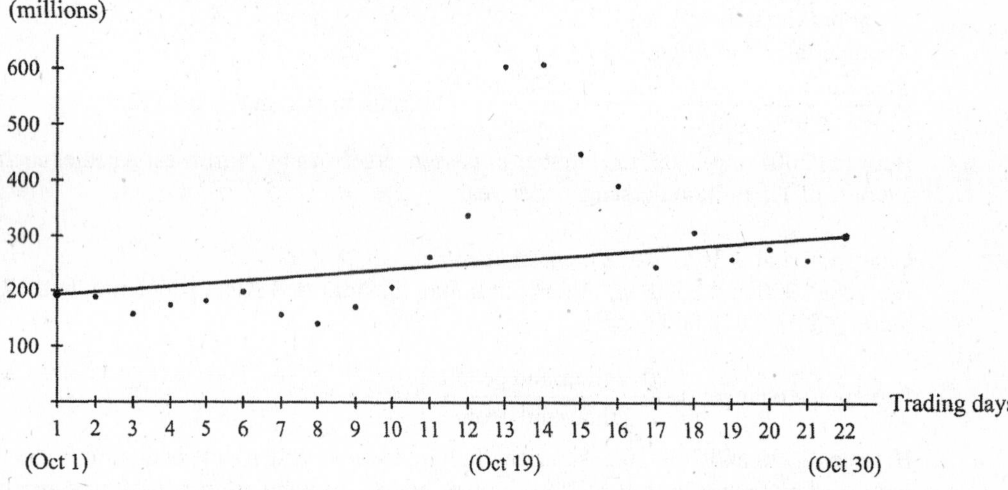

c. The volume of shares traded on a given day in October 1987 stayed near (or below) 200 million shares until mid-October when it spiked near 600 million shares. It then declined to near 300 million shares for the remainder of the month. The average rate of change ignores the spike during "October Madness."

11. a. $\text{Average rate of change} = \frac{4.25 - 3.25}{2001 - 1996} = \$0.2 \text{ billion per year}$

Between 1996 and 2001, sales at Kelly Services, Inc. increased by an average of $0.2 billion per year.

b. $\text{Percentage change} = \frac{4.25 - 3.25}{3.25} \cdot 100\% = 30.7\%$

Between 1996 and 2001, sales at Kelly Services, Inc. increased by 30.7%.

13. a. Because a scatter plot of the data is concave down, a quadratic model is best.
$P(x) = -0.037x^2 + 25.529x - 527.143$ thousand dollars monthly profit for an airline from a roundtrip flight from Denver to Chicago, where x dollars is the cost of a roundtrip ticket and $200 \le x \le 450$.

b. $P(200) \approx 3081.429$ thousand dollars; $P(350) \approx 3822.857$ thousand dollars

$$\text{Average rate of change} \approx \frac{3822.857 - 3081.429 \text{ thousand dollars of profit}}{350 - 200 \text{ dollars of ticket price}}$$
$$\approx 4.943 \text{ thousand dollars of profit per dollar of ticket price}$$

c. $P(350) \approx 3822.857$ thousand dollars; $P(450) \approx 3381.429$ thousand dollars

$$\text{Average rate of change} \approx \frac{3381.429 - 3822.857 \text{ thousand dollars of profit}}{450 - 350 \text{ dollars of ticket price}}$$
$$\approx -4.414 \text{ thousand dollars of profit per dollar of ticket price}$$

15. a. Average rate of change: $\frac{11.8 \text{ years} - 68.3 \text{ years}}{70} \approx -0.81$ year per year (year of life expectancy per year of age)

b. Average rate of change between ages 10 and 20: $\frac{50 \text{ years} - 59.6 \text{ years}}{10} \approx -0.96$ year per year (year of life expectancy per year of age)

Average rate of change between ages 20 and 30: $\frac{41.5 \text{ years} - 50 \text{ years}}{10} \approx -0.89$ year per year (year of life expectancy per year of age)

Life expectancy decreases with increasing age, but the magnitude of the rate of decrease gets smaller as a male gets older.

17. a. $p(55) - p(40) \approx 31.183 - 21.218$ million people ≈ 9.965 million people

Percentage change: $\left[\frac{p(55) - p(40)}{p(40)}\right]100\% \approx \left(\frac{9.965 \text{ million people}}{21.218 \text{ million people}}\right)100\% \approx 47\%$

b. $\frac{p(85) - p(83)}{85 - 83} \approx \frac{67.350\text{-}63.980 \text{ million people}}{2 \text{ years}} \approx 1.685$ million people per year

19. 1996: $s(6) \approx 16.244\%$ 1999: $s(11) \approx 95.593\%$

Percentage change $= \left(\frac{95.593 - 16.244}{16.244}\right)100\% \approx 488.5\%$

21. a. i. $\frac{y(3) - y(1)}{3 - 1} = \frac{13 - 7}{2} = \frac{6}{2} = 3$

ii. $\frac{y(6) - y(3)}{6 - 3} = \frac{22 - 13}{3} = \frac{9}{3} = 3$

iii. $\frac{y(10) - y(6)}{10 - 6} = \frac{34 - 22}{4} = \frac{12}{4} = 3$

b. i. $\left[\frac{y(3)-y(1)}{y(1)}\right]100\% = \left(\frac{6}{7}\right)100\% \approx 85.7\%$

ii. $\left[\frac{y(5)-y(3)}{y(3)}\right]100\% = \left(\frac{6}{13}\right)100\% \approx 46.2\%$

iii. $\left[\frac{y(7)-y(5)}{y(5)}\right]100\% = \left(\frac{6}{19}\right)100\% \approx 31.6\%$

c. The average rate of change of any linear function over any interval will be constant because the slope of a line (and, therefore, of any secant line) is constant. The percentage change is not constant.

23. a. The balance increased by \$1908.80 – \$1489.55 = \$419.25.

b. Average rate of change: $\frac{\$419.25}{4 \text{ years}} \approx \$104.81$ per year

Between the end of year 1 and the end of year 5, the balance increased at an average rate of \$104.81/ year.

c. There are no data available for the balance in the account at the middle of the fourth year. Without a model, there is no way to estimate the average rate of change in the balance from the middle to the end of the fourth year.

d. The amount in the account can be modeled as $B(x) = 1400(1.064^x)$ dollars, where x is the number of years since the initial investment, $1 \le x \le 5$.

Amount in middle of year 4: $B(3.5) = \$1739.28$
Amount at end of year 4: $B(4) = \$1794.04$

Average rate of change $= \frac{\$1794.04 - \$1739.28}{\frac{1}{2} \text{ year}} = \$109.52$ per year

25. a. $\text{APR} = 1.5\frac{\%}{\text{month}} \cdot 12 \text{ months} = 18\%$

b. $A(t) = P\left(1+\frac{0.18}{12}\right)^{12t} \approx P(1.195618)^t$

$\text{APY} = (1.195618-1)100\% \approx 19.562\%$

27. a. $2P = P\left(1+\frac{r}{n}\right)^{nt}$

$$2 = \left(1+\frac{0.063}{12}\right)^{12t}$$

$$\ln 2 = 12t \ln\left(1+\frac{0.063}{12}\right)$$

$$t = \frac{\ln 2}{12 \ln\left(1+\frac{0.063}{12}\right)} \approx 11.03$$

It will take just over 11 years (that is, 11 years 1 month).

b. $2P = Pe^{rt}$

$2 = e^{0.08t}$

$\ln 2 = 0.08t$

$t = \dfrac{\ln 2}{0.08} \approx 8.66$

It will take approximately 8.66 years (8 years 8 months).

c. $2P = P\left(1+\dfrac{r}{n}\right)^{nt}$

$2 = \left(1+\dfrac{0.0685}{4}\right)^{4t}$

$\ln 2 = 4t\ln\left(1+\dfrac{0.0685}{4}\right)$

$t = \dfrac{\ln 2}{4\ln\left(1+\frac{0.0685}{4}\right)} \approx 10.21$

It will take 10 years 3 months.

29. a. $A = \left(1+\frac{1}{n}\right)^n$

b,c.

Compounding	n	Amount
yearly	1	\$2.00
semiannually	2	\$2.25
quarterly	4	\$2.44
monthly	12	\$2.61
weekly	52	\$2.69
daily	365	\$2.71
every hour	8760	\$2.72
every minute	525,600	\$2.72
every second	31,636,000	\$2.72

d. \$2.72

e. $\lim_{n\to\infty}\left(1+\frac{1}{n}\right)^n \approx 2.72$

31. *One possible answer:* Change is simply a report of the difference in two output values so that the magnitude of the change may be considered. Percentage change is a report of the difference in two output values so that the relative magnitude of the change may be considered. Average rate of change is a report of the difference in two output values in a way that considers the associated spread of the input values.

Section 2.2 Instantaneous Rates of Change

1. *One possible set of answers:*

a. A continuous graph or model is defined for all possible input values on an interval. A continuous model with discrete interpretation has meaning for only certain input values on an interval. A continuous graph can be drawn without lifting the pencil from the paper. A discrete graph is a scatter plot. A continuous model or graph can be used to find average or instantaneous rates of change. Discrete data or a scatter plot can be used to find average rates of change.

b. An average rate of change is a slope between two points. An instantaneous rate of change is the slope at a single point on a graph.

c. A secant line connects two points on a graph. A tangent line touches the graph at a point and is tilted the same way the graph is tilted at that point.

3. Average rates of change are slopes of secant lines. Instantaneous rates of change are slopes of tangent lines.

5. Average speed $= \dfrac{19-0 \text{ miles}}{17 \text{ minutes}} \cdot \dfrac{60 \text{ minutes}}{\text{hour}} \approx 67.1 \text{ mph}$

7. a. The slope is positive at A, negative at B and E, and zero at C and D.

b. The graph is slightly steeper at point B than at point A.

9.

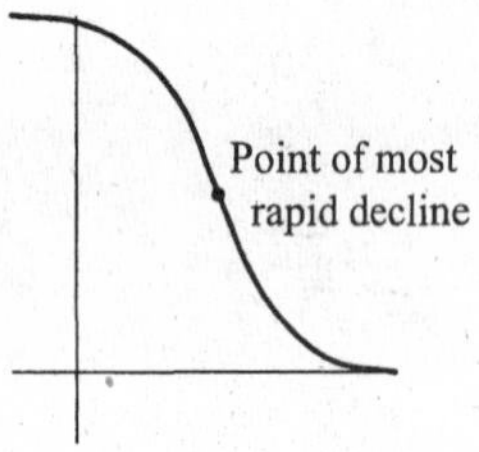

Point of most rapid decline

11. The graph shows a decreasing, linear function. The slope is a constant, negative number.

13. The graph shows an increasing, logistic function. The slope is always positive. Moving from left to right, the slope starts close to zero, increases, and then decreases toward zero.

15. The lines drawn at A and C are not tangent lines because they don't follow the tilt of the curve.

17.

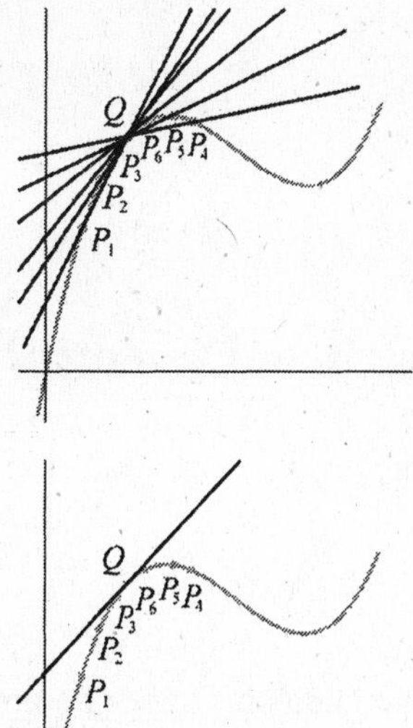

19. a,b. *A*: concave down, tangent line lies above the curve
B: inflection point, tangent line lies below the curve on the left, above the curve on the right
C: inflection point, tangent line lies above the curve on the left, below the curve on the right. **Note: the answer key in the text is in error concerning point *C*.**
D: concave up, tangent line lies below the curve.

c.

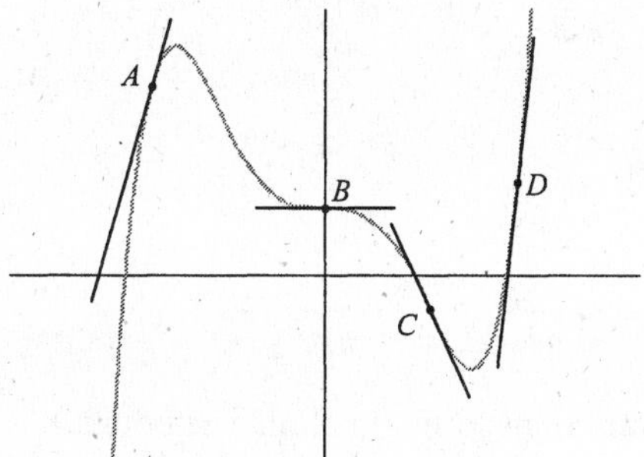

d. *A*, *D*: Positive slope
C: Negative slope (inflection point)
B: Zero slope (inflection point)

21.

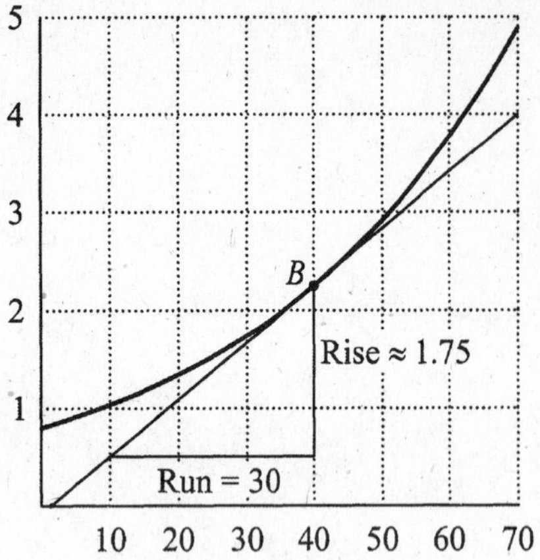

Slope $= \dfrac{\text{rise}}{\text{run}} \approx \dfrac{1.75}{30} \approx 0.58$ (Estimates may vary slightly depending on choice of points.)

Note: the answer calculations in the back of the text used a different choice of points but arrived at exactly the same estimate.

23.

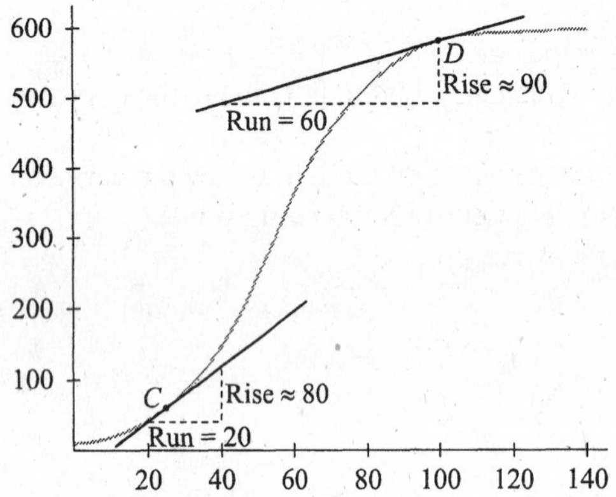

Slope at $C = \dfrac{\text{rise}}{\text{run}} \approx \dfrac{80}{20} = 4$

Slope at $D = \dfrac{\text{rise}}{\text{run}} \approx \dfrac{90}{60} = 1.5$ (Estimates may vary depending on choice of points.)

25. a. Million subscribers per year

b. In 2000, the number of subscribers was increasing by 23.1 million subscribers per year.

c. 23.1 million subscribers per year

d. 23.1 million subscribers per year

27. a, b. A: 1.3 mm per day per °C

B: 5.9 mm per day per °C

C: –4.2 mm per day per °C

c. The growth rate is increasing by 5.9 mm per day per °C.

d. The slope of the tangent line at 32°C is –4.2 mm per day per °C.

e. At 17°C, the instantaneous rate of change is 1.3 mm per day per °C.

29. a. Declination of sun

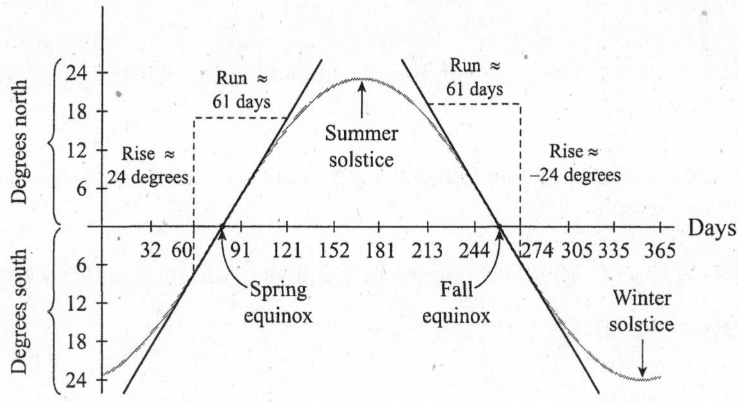

The slope at the solstices is zero.

b. The steepest points on the graph are those where the graph crosses the horizontal axis. The slopes are estimated as

$$\frac{24 \text{ degrees}}{61 \text{ days}} \approx 0.4 \text{ degree per day and } \frac{-24 \text{ degrees}}{61 \text{ days}} \approx -0.4 \text{ degree per day}$$

A negative slope indicates that the sun is moving from north to south.

31. a. Because the model is linear, the line to be sketched is the same as the model itself. From the equation, its slope is approximately 2.37 million people per year.

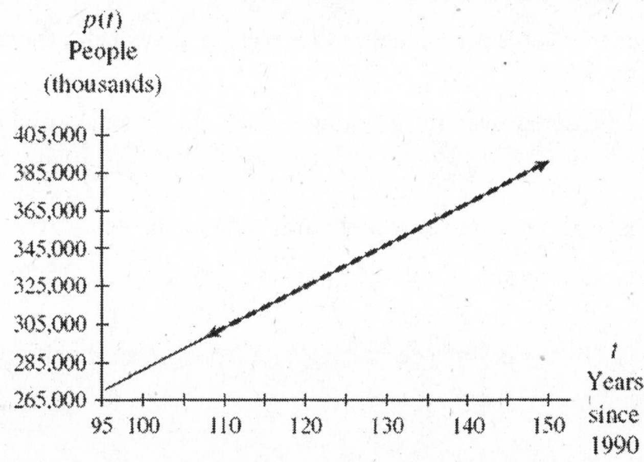

b. Any line tangent to the graph of $p(t)$ coincides with the graph itself.

c. Any line tangent to this graph has a slope of approximately 2.37 million people per year.

d. The slope of the graph at every point will be 2.37 million people per year.

e. The instantaneous rate of change is 2.37 million people per year.

33.

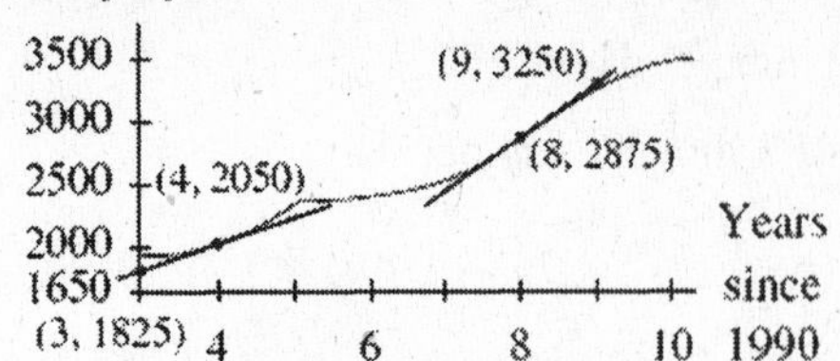

a. $\frac{2050-1825}{4-3}=225$: In 1994, the number of employees was increasing by approximately 225 employees per year.

b. The slope of the graph cannot be found at $x = 5$ because a tangent cannot be drawn at a sharp point on a graph.

c. $\frac{3250-2875}{9-8}=375$: In 1998, the number of employees was increasing by approximately 375 employees per year.

35. *One possible answer:* The line tangent to a graph at a point P is the limiting position of secant lines through P and nearby points on the graph.

Section 2.3 Derivative Notation and Numerical Estimates

1. a. Because $P(t)$ is measured in miles and t is measured in hours, the units are miles per hour.

b. Speed or velocity

3. a. No, the number of words per minute cannot be negative.

b. Because $w(t)$ is measured in words per minute and t is measured in weeks, the units are words per minute per week.

c. The student's typing speed could actually be getting worse, which would mean that $W'(t)$ is negative.

5. a. When the ticket price is \$65, the airline's weekly profit is \$15,000.

b. When the ticket price is \$65, the airline's weekly profit is increasing by \$1500 per dollar of ticket price.

c. When the ticket price is \$90, the profit is decreasing \$2000 per dollar of ticket price.

7.

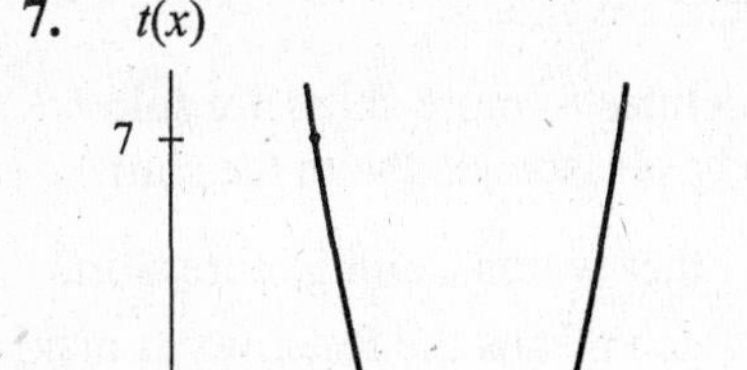

9. a. At the beginning of the diet you weigh 167 pounds.

b. After 12 weeks of dieting your weight is 142 pounds.

c. After 1 week of dieting your weight is decreasing by 2 pounds per week.

d. After 9 weeks of dieting your weight is decreasing by 1 pound per week.

e. After 12 weeks of dieting your weight is neither increasing nor decreasing.

f. After 15 weeks of dieting you are gaining weight at a rate of a fourth of a pound per week.

g.

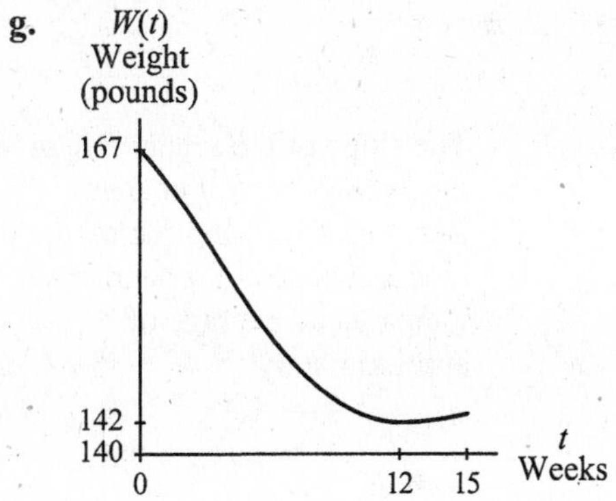

11. Because $21 + 12 = 32$ and $12 + 10(0.6) = 18$, we know that the following points are on the graph: (1940, 4), (1970, 12), (2000, 33), and (1980, 18). We also know the graph is concave up between 1940 and 1990 and concave down between 1990 and 1995. *One possible graph:*

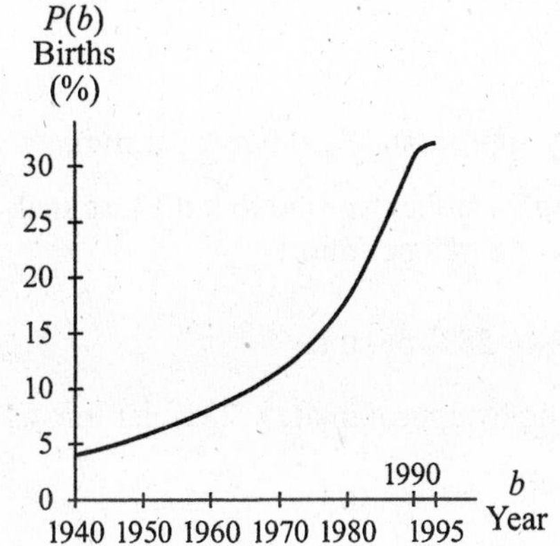

13. a. It is possible for profit to be negative if costs are more than revenue.

b. It is possible for the derivative to be negative if profit declines as more shirts are sold (because the price is so low, the revenue is less than the cost associated with the shirt.)

c. If $P'(200) = -1.5$, the fraternity's profit is declining. In other words, selling more shirts would result in less profit. Profit may still be positive (which means the fraternity is making money), but the negative rate of change indicates they are not making the most profit possible (they could make more money by selling fewer shirts).

15. a. Because $D(r)$ is measured in years and r is measured in percentage points, the units on $\frac{dD}{dr}$ are years per percentage point.

b. As the rate of return increases, the time it takes the investment to double decreases.

c. i. When the interest rate is 9%, it takes 7.7 years for the investment to double.

ii. When the interest rate is 5%, the doubling time is decreasing by 2.77 years per percentage point. A one-percentage-point increase in the interest rate will decrease the doubling time by approximately 2.8 years.

iii. When the interest rate is 12%, the doubling time is decreasing by 0.48 year per percentage point. A one-percentage-point increase in the interest rate will decrease the doubling time by approximately half of a year.

17. a,b.

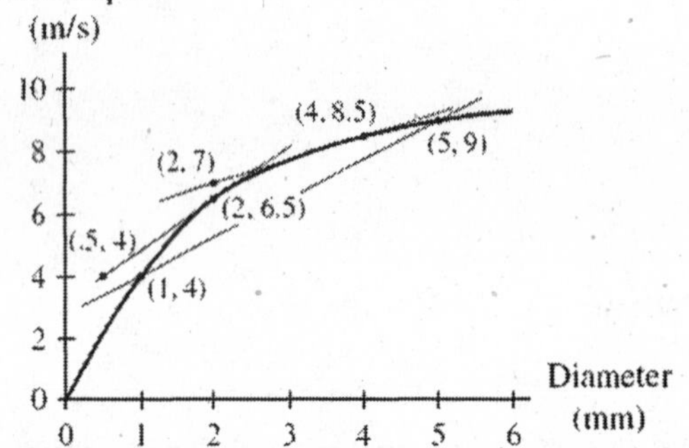

The slope of the secant line gives the average rate of change. Between 1 mm and 5 mm, the terminal speed of a raindrop increases an average of approximately

$$\frac{5 \text{ m/s}}{4 \text{ mm}} = 1.25 \text{ m/s per mm.}$$

The slope of the tangent line gives the instantaneous rate of change of the terminal speed of a 4 mm raindrop.

c. $\text{Slope} = \frac{\text{rise}}{\text{run}} \approx \frac{1.25 \text{ m/s}}{2 \text{ mm}} = 0.625 \text{ m/s per mm}$

A 4 mm raindrop's terminal speed is increasing by approximately 0.6 m/s per mm.

d. By sketching a tangent line at 2 mm and estimating its slope, we find that the terminal speed of a raindrop is increasing by approximately 1.8 m/s per mm.

e. $\text{Percentage rate of change} = \frac{1.8 \text{ m/s per mm}}{6.4 \text{ m/s}} \cdot 100\% \approx 28\% \text{ per mm}$

The terminal speed of a 2 mm raindrop is increasing by approximately 28% per mm as the diameter increases.

19. a.

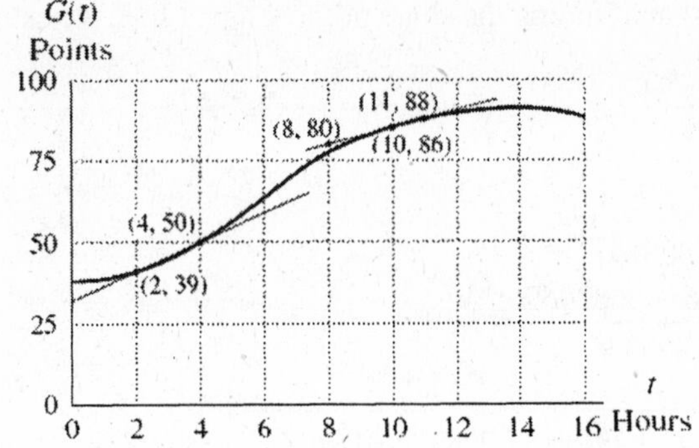

Slope at 4 hours $\approx \frac{11 \text{ points}}{2 \text{ hours}}$ = 5.5 points per hour

Slope at 11 hours $\approx \frac{8 \text{ points}}{3 \text{ hours}}$ = 2.7 points per hour

(Estimates may vary slightly.)

b. Estimate the two points on the graph: (4, 50) and (10, 86).

Average rate of change $\approx \frac{86 \text{ points} - 50 \text{ points}}{10 \text{ hours} - 4 \text{ hours}} = \frac{36 \text{ points}}{6 \text{ hours}} = 6$ points per hour

As the number of hours that you study increases from 4 to 10 hours, your expected grade on the calculus test increases by an average of 6 points per hour.

c. Percentage rate of change $\approx \frac{6 \text{ points per hour}}{50 \text{ points}} \cdot 100\% = 12\%$ per hour. When you have studied for 4 hours, your expected grade on the calculus test is increasing by 12% per hour.

d. $G(4.6) \approx G(4) + 0.6 \cdot G'(4) = 50 + 0.6 \cdot 5.5 = 53.3$ points

21. a.

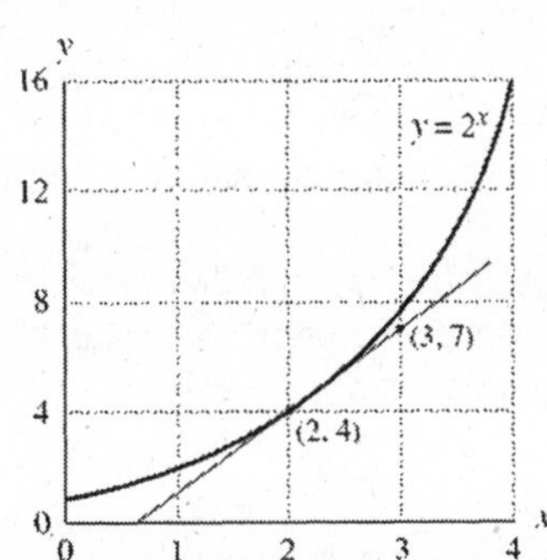

b. Sample calculation: **Note: the answers in the back of the text are incorrect.**

Point at $x = 2$: (2, 4)   Point at $x = 1.9$: $(1.9, 2^{1.9}) \approx (1.9, 3.732132)$

Secant line slope $\approx \frac{3.732132 - 4}{1.9 - 2} = \frac{-0.267868}{-0.1} = 2.67868$

Input of close point on left	Slope	Input of close point on right	Slope
1.9	2.67868	2.1	2.87094
1.99	2.76300	2.01	2.78222
1.999	2.77163	2.001	2.77355
1.9999	2.77249	2.0001	2.77268
1.99999	2.77258	2.00001	2.77260
Limit ≈ 2.77		**Limit ≈ 2.77**	

The slope of the line tangent to $y = 2^x$ at $x = 2$ is approximately 2.77.

23. a. Sketching a tangent line at $t = 4$ and finding the slope of the tangent line, we estimate that $\frac{dP}{dt} \approx 3$ million passengers per year.

b. Sample calculation:

Point at $x = 4$: (4, 83.16)

Point at $x = 3.9$: (3.9, 82.82069)

Secant line slope $= \frac{83.16 - 82.82069}{4 - 3.9} = 3.3931$

Input of close point on left	Slope	Input of close point on right	Slope
3.9	3.3931	4.1	3.2811
3.99	3.3456	4.01	3.3344
3.999	3.3406	4.001	3.3394
3.9999	3.3401	4.0001	3.3400
3.99999	3.3400	4.00001	3.3400
Limit ≈ 3.34		**Limit ≈ 3.34**	

$P'(4) \approx 3.34$ million passengers per year

In 2004, the number of passengers going through the Atlanta International Airport each year was increasing by approximately 3.34 million passengers per year.

c. $\text{PROC} = \frac{P'(4)}{P(4)} \cdot 100\% \approx \frac{3.34 \text{ million passengers per year}}{83.16 \text{ million passengers}} \cdot 100\% \approx 4.016\%$ per year

In 2004, the number of passengers going through the Atlanta International Airport each year was increasing by approximately 4% per year.

d. *One possible answer:* Using numbers based on an equation is more accurate, but there are many situations when the equation of a graph is not provided.

25. a. Sample calculation:

Point at $x = 13$: $(13,\ 0.181(13)^2 - 8.463(13) + 147.376) = (13, 67.946)$

Point at $x = 12.9$: $(12.9,\ 0.181(12.9)^2 - 8.463(12.9) + 147.376) = (12.9, 68.32351)$

Secant line slope $\approx \frac{68.32351 - 67.946}{12.9 - 13} = -3.7751$ seconds per year of age

Input of close point on left	Slope	Input of close point on right	Slope
12.9	–3.7751	13.1	–3.7389
12.99	–3.75881	13.01	–3.75519
12.999	–3.757181	13.001	–3.756819
12.9999	–3.7570181	13.0001	–3.7569819
12.99999	–3.75700181	13.00001	–3.75699819
Limit from either direction ≈ –3.757 seconds per year of age			

b. $\text{PROC} = \frac{T'(13)}{T(13)} \cdot 100\% \approx \frac{\text{-3.757 seconds per year}}{\text{67.946 seconds}} \cdot 100\% \approx -5.529$ percent per year

c. The swimmer's time is decreasing. Thus the swimmer is swimming faster, and the time is improving.

27. a. $P(x)$ and $C(P)$ can be combined by function composition to create the profit function. $A(x) = C(P(x)) = \frac{1.02^x}{1.5786}$ is the profit in American dollars from the sale of x mountain bikes.

b. Canadian: $C(400) \approx \$2754.66$

American: $A(400) \approx \$1745.00$

c. We can estimate $A'(400) \approx \$34.56$ per mountain bike sold

Input of close point on left	Slope	Input of close point on right	Slope
399.9	34.521	400.1	34.590
399.99	34.552	400.01	34.559
399.999	34.555	400.001	34.556
399.9999	34.556	400.0001	34.556
399.99999	34.556	400.00001	34.556
Limit from either direction ≈ $34.56 per mountain bike			

29. *One possible answer:* The *percentage change* gives the relative amount of change in the output from an initial input value to a second input value. The *percentage rate of change* is a relative measure of the rate of change at a particular input value in comparison to the output value at that point.

31. *One possible answer:* Finding the rate of change *numerically* using numbers found with an equation is both more accurate and more time-consuming than drawing a *tangent line* to a graph and estimating the slope of that tangent line. However, there are many situations when the equation of a graph is not provided and the rate of change must be estimated graphically.

Section 2.4 Algebraically Finding Slopes

1.

$x \to 2^-$	$\frac{x^3}{x-2}$	$x \to 2^+$	$\frac{x^3}{x-2}$
1.9	–68.59	2.1	92.61
1.99	–788.0599	2.01	812.0601
1.999	–7988.00599	2.001	8012.006001
1.9999	–79988.0006	2.0001	80012.0006

$\lim\limits_{x\to 2^-} \frac{x^3}{x-2} \to -\infty$; $\lim\limits_{x\to 2^+} \frac{x^3}{x-2} \to \infty$; $\lim\limits_{x\to 2} \frac{x^3}{x-2}$ does not exist

3.

$x \to 0^-$	$\frac{-2x^3+7x}{x}$	$x \to 0^+$	$\frac{-2x^3+7x}{x}$
–0.1	6.98	0.1	6.98
–0.01	6.9998	0.01	6.9998
–0.001	6.999998	0.001	6.999998
–0.0001	6.99999998	0.0001	6.99999998

$\lim_{x\to 0^-}\frac{-2x^3+7x}{x}=7;\ \lim_{x\to 0^+}\frac{-2x^3+7x}{x}=7;\ \lim_{x\to 0}\frac{-2x^3+7x}{x}=7$

5. $f'(2)=\lim_{x\to 2}\frac{f(x)-f(2)}{x-2}=\lim_{x\to 2}\frac{4x^2-16}{x-2}=16$

$x \to 2^-$	$\frac{4x^2-16}{x-2}$	$x \to 2^+$	$\frac{4x^2-16}{x-2}$
1.9	15.6	2.1	16.4
1.99	15.96	2.01	16.04
1.999	15.996	2.001	16.004
1.9999	15.9996	2.0001	16.0004

7. $g'(4)=\lim_{t\to 4}\frac{g(t)-g(4)}{t-4}=\lim_{t\to 4}\frac{\left(-6t^2+7\right)-\left(-89\right)}{t-4}=\lim_{t\to 4}\frac{-6t^2+96}{t-4}=-48$

$t \to 4^-$	$\frac{-6t^2+96}{t-4}$	$t \to 4^+$	$\frac{-6t^2+96}{t-4}$
3.9	–47.4	4.1	–48.6
3.99	–47.94	4.01	–48.06
3.999	–47.994	4.001	–48.006
3.9999	–47.9994	4.0001	–48.0006

Or using direct substitution:

$$\lim_{t\to 4}\frac{-6t^2+96}{t-4}=\lim_{t\to 4}\frac{-6(t-4)(t+4)}{t-4}=\lim_{t\to 4}-6(t+4)=-6(4+4)=-48$$

9. Point: $(x, f(x)) \to (x,\ 3x-2)$

Close Point: $(x+h, f(x+h)) \to (x+h,\ 3(x+h)-2) \to (x+h,\ 3x+3h-2)$

Slope of the Secant: $\frac{3x+3h-2-(3x-2)}{x+h-x}=\frac{3x+3h-2-3x+2}{h}=\frac{3h}{h}$

Limit of the Slope of the Secant: $\frac{dy}{dx}=\lim_{h\to 0}\frac{3h}{h}=\lim_{h\to 0}3=3$

11. Point: $(x, f(x)) \rightarrow (x,\ 3x^2)$

Close Point: $(x+h, f(x+h)) \rightarrow (x+h,\ 3(x+h)^2) \rightarrow (x+h,\ 3(x+h)(x+h)) \rightarrow$
$(x+h,\ 3(x^2+2xh+h^2)) \rightarrow (x+h,\ 3x^2+6xh+3h^2)$

Slope of the Secant: $\dfrac{(3x^2+6xh+3h^2)-3x^2}{x+h-x} = \dfrac{6xh+3h^2}{h} = \dfrac{h(6x+3h)}{h}$

Limit of the Slope of the Secant: $f'(x) = \lim\limits_{h\to 0}\dfrac{h(6x+3h)}{h} = \lim\limits_{h\to 0}(6x+3h) = 6x+3(0) = 6x$

13. Point: $(x, f(x)) \rightarrow (x,\ x^3)$

Close Point: $(x+h, f(x+h)) \rightarrow (x+h,\ (x+h)^3) \rightarrow (x+h,\ x^3+3x^2h+3xh^2+h^3)$

Slope of the Secant: $\dfrac{(x^3+3x^2h+3xh^2+h^3)-x^3}{x+h-x} = \dfrac{3x^2h+3xh^2+h^3}{h} = \dfrac{h(3x^2+3xh+h^2)}{h}$

Limit of the Slope of the Secant:

$$\frac{dy}{dx} = \lim_{h\to 0}\frac{h(3x^2+3xh+h^2)}{h} = \lim_{h\to 0}(3x^2+3xh+h^2) = 3x^2+3x(0)+0^2 = 3x^2$$

15. a. $T(13) = 67.946$ seconds

b. $T(13+h) = 0.181(13+h)^2 - 8.463(13+h) + 147.376 = 0.181h^2 - 3.757h + 67.946$

c. $\dfrac{T(13+h)-T(13)}{13+h-13} = \dfrac{0.181h^2 - 3.757h + 67.946 - 67.946}{h} = \dfrac{0.181h^2 - 3.757h}{h}$

d. $\lim\limits_{h\to 0}\dfrac{0.181h^2 - 3.757h}{h} = \lim\limits_{h\to 0}(0.181h - 3.757) = 0.181(0) - 3.757$
$= -3.757$ seconds per year of age

The swim time for a 13-year-old is decreasing by 3.757 seconds per year of age. This tells us that as a 13-year-old athlete gets older, the athlete's swim time improves.

17. a. $f(3) = 2.052$ billion gallons

b. $f(3+h) = -0.042(3+h)^2 + 0.18(3+h) + 1.89 = 2.052 - 0.072h - 0.042h^2$ billion gallons

c. $\dfrac{f(3+h)-f(3)}{3+h-3} = \dfrac{(2.052-0.072h-0.042h^2)-2.052}{h} = \dfrac{-0.072h-0.042h^2}{h}$
$= \dfrac{h(-0.072-0.042h)}{h}$ billion gallons per year

d. $\lim_{h\to 0}\dfrac{h(-0.072-0.042h)}{h}=\lim_{h\to 0}(-0.072-0.042h)=-0.072-0.042(0)=-0.072$

In 2001, the amount of fuel Northwest Airlines consumed each year was decreasing by 72 million gallons per year.

19. a. Point: $\left(t,\ -16t^2+100\right)$

Close Point: $\left(t+h,\ -16(t+h)^2+100\right)\rightarrow\left(t+h,\ -16(t^2+2th+h^2)+100\right)$

$\rightarrow\left(t+h,\ -16t^2-32th-16h^2+100\right)$

Slope of the Secant:

$$\frac{\left(-16t^2-32th-16h^2+100\right)-\left(-16t^2+100\right)}{t+h-t}=\frac{-16h^2-32th}{h}=\frac{h\left(-16h-32t\right)}{h}.$$

Limit of the Slope of the Secant:

$$\frac{dH}{dt}=\lim_{t\to 0}\frac{h\left(-16h-32t\right)}{h}=\lim_{t\to 0}\left(-16h-32t\right)=\left(-16(0)-32t\right)=-32t \text{ feet per second}$$

is the speed of a falling object t seconds after the object begins falling (given that the object has not reached the ground).

b. $\left.\dfrac{dH}{dt}\right|_{t=1}=-32(1)=-32$ feet per second

After 1 second, the object is falling at a speed of 32 feet per second.

21. a. The number of drivers of age a years in 1997 can be modeled as $D(a)=-0.045a^2+1.774a-16.064$ million drivers, $16\le a\le 21$.

b.

$$D'(a) = \lim_{h\to 0}\frac{D(a+h)-D(a)}{h}$$

$$= \lim_{h\to 0}\frac{\left(-.045(a+h)^2+1.774(a+h)-16.064\right)-\left(-0.045a^2+1.774a-16.064\right)}{h}$$

$$= \lim_{h\to 0}\frac{\left(-0.045a^2-0.09ah-0.045h^2+1.774a+1.774h-16.064\right)+0.045a^2-1.774a+16.064}{h}$$

$$= \lim_{h\to 0}\frac{-0.09ah-0.045h^2+1.774h}{h}$$

$$= \lim_{h\to 0}\frac{h\left(-0.09a-0.045h+1.774\right)}{h}$$

$$= \lim_{h\to 0}\left(-.09a-.045h+1.774\right)$$

$$= -0.09a-0.045(0)+1.774$$

$$= -0.09a+1.774$$

Thus $D'(a) = -0.09a+1.774$ million drivers per year of age where a is the age in years

c. $D'(20) = -0.026$ million drivers per year of age. For 20-year-olds, the number of licensed drivers in 1997 was decreasing by 26,000 drivers per year of age.

23. *One possible answer:* Finding a slope graphically is the only approach if all that is available is a graph of the function without an accompanying equation. Finding a slope graphically may be appropriate if all that is needed is a quick approximation of the rate of change at a point. If a more precise determination of the rate of change at one single point is needed and an equation is available, it may be appropriate to find the slope at that one point numerically, using a table of limiting values of the slopes of increasingly close secant lines. If an equation is available and the rate of change at several different points is needed, it might be appropriate to use an algebraic method to find a formula for the slope. A final consideration when choosing between the algebraic method and the numerical method to find slope is the difficulty involved in using the algebraic method. We are generally limited to using the *algebraic* method for linear, quadratic, or cubic functions.

Chapter 2 Concept Review

1. a. i. A, B, C **ii.** E **iii.** D

b. The graph is steeper at B than it is at A, C, or D.

c. Below: C, D, E
Above: A
At B: above to the left of B, below to the right of B

d.

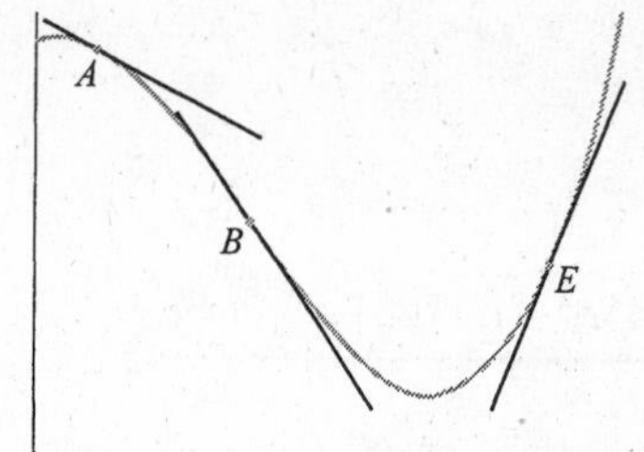

2. **a.** Feet per second per second or feet per second squared. This is acceleration.
   b. The speed of the roller coaster increased after point D.
   c. The roller coaster's speed was slowest at point D.
   d. The roller coaster was slowing down most rapidly at point B.

3. **a.** The number of states associated with the national P.T.A. association grew by an average of 1.125 states per year between 1915 and 1931. (calculate the slope of the secant line between 1915 and 1931; estimates may vary)

   b. The number of states associated with the national P.T.A. association grew by approximately 60% between 1915 and 1931. (Estimates may vary.)

   c. The number of states associated with the national P.T.A. association grew by an average of 1 state per year between 1923 and 1927. (calculate the slope of the secant line between 1923 and 1927; estimates may vary)

4. **a.** Solve the following equation for P: $25{,}000 = P\left(1+\dfrac{0.075}{12}\right)^{(12\cdot 15)}$

   Approximately $8144.78

   b. $A(15) = 25{,}000\left(1+\dfrac{0.065}{4}\right)^{(4\cdot 15)} \approx \$65{,}761.77$

   c. $A(t) = 25000\left(1+\dfrac{0.065}{4}\right)^{4t}$ dollars in an account after t years when interest is compounded quarterly at 6.5%

   d. $\text{AROC} = \dfrac{A(10)-A(5)}{10-5} \approx 2625.695$ dollars per year

   $\text{Percentage Change} = \dfrac{A(10)-A(5)}{A(5)}100\% \approx 38.042\%$

5. **a,b.** The slope of the secant line gives the average rate of change between 1993 and 1997.

   The slope of the tangent line gives the instantaneous rate of change in 1998.

   c. Between 1993 and 1997, the number of Dell employees increased at an average rate of 1500 employees per year. (slope of the secant line; estimates may vary)

d. $E'(1998) \approx \frac{15000-8000}{1998-1996} = 3500$ employees per year; In 1998 the number of employees at Dell Computer Corp. was increasing by approximately 3500 employees per year. (slope of the tangent line; estimates may vary)

$\frac{3500 \text{ employees/year}}{15000 \text{ employees}} \cdot 100\% \approx 23.3\ \%$ per year ; In 1998 the number of employees at Dell Computer Corp. was increasing by approximately 23.3% per year.

6. a. An average 22-year-old athlete can swim 100 meters in 49 seconds.
The time required for an average 22-year-old athlete to swim 100 meters free style is decreasing by approximately half a second per year of age.

b. A negative rate of change indicates that the average swimmer's time improves as age increases.

7. a. $R(x) = -0.051x^2 + 0.884x + 4.793$ billion dollars passenger revenue for Northwest Airlines between 1991 and 2003, where x is the number of years since 1990.

b. Approximately –0.14 billion dollars per year

Close point	Slope of the secant line between point and $x = 10$
9.9	–0.1336
9.99	–0.1382
9.999	–0.1386
9.9999	–0.1387
10.1	–0.1488
10.01	–0.1392
10.001	–0.1387
10.0001	–0.1387

c. In 2001, passenger revenue for Northwest Airlines was decreasing by approximately $140 million per year.

d. $R'(x) = -0.102x + 0.884$ billion dollars per year is how quickly passenger revenue for Northwest Airlines was changing between 1991 and 2003, where x is the number of years since 1990; $R'(11) \approx -\$0.241$ billion per year (found using the full model)

8. Find a point: $(x,\ 7x+3)$; find a close point: $(x+h,\ 7(x+h)+3)$; find a formula for the slope of the secant line between the two points and simplify completely:

Slope of the secant $= \dfrac{7x+7h+3-(7x+3)}{x+h-x} = 7$; find the limit of the slope of the secant as the point and the close point become closer together: $\lim_{h\to 0} 7 = 7$.

Chapter 3
Determining Change: Derivatives

Section 3.1 Drawing Rate-of-Change Graphs

1. The slopes are negative to the left of A and positive to the right of A. The slope is zero at A.

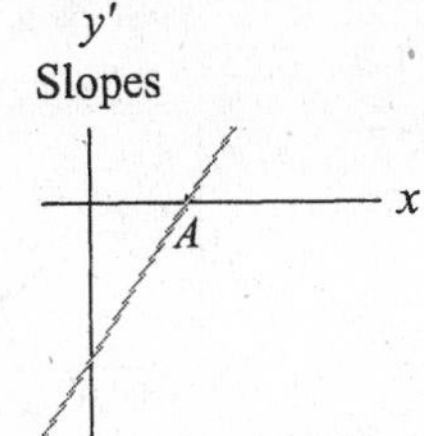

3. The slopes are positive everywhere, near zero to the left of zero, and increasingly positive to the right of zero.

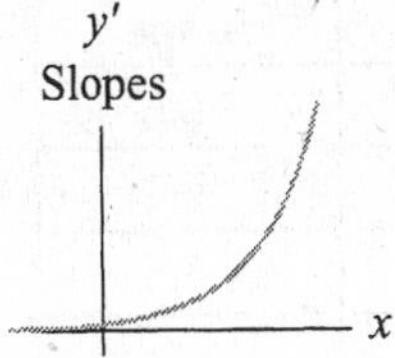

5. The slope is zero everywhere.

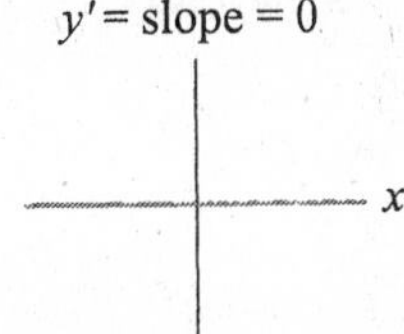

7. The slopes are negative everywhere. The magnitude is large close to $x = 0$ and is near zero to the far right.

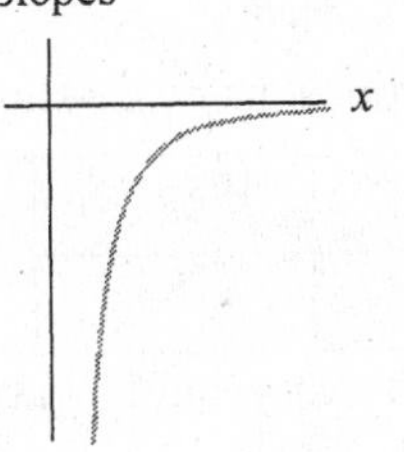

9. The slopes are negative to the left and right of A. The slope appears to be zero at A.

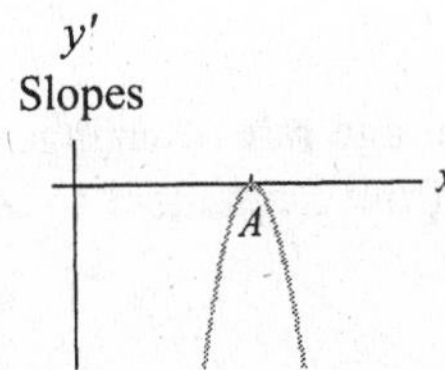

11. a.

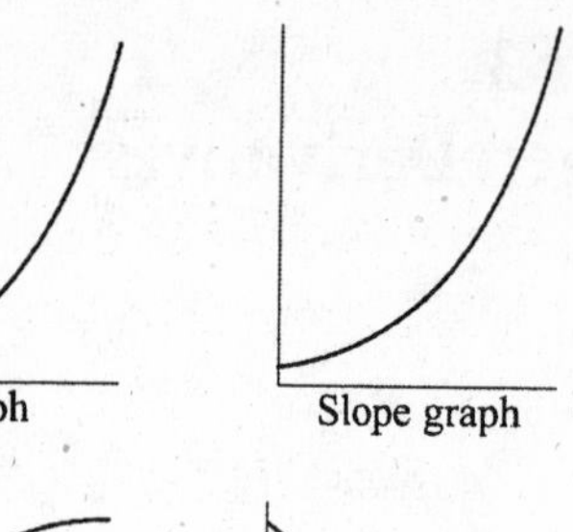

Graph Slope graph

b.

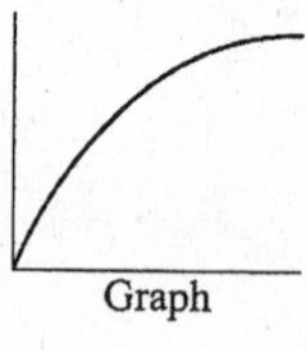

Graph Slope graph

13. a, b.

Year	Slope
1991	−6.6
1993	−6.2
1997	−2.5
1999	1.0
2000	5.4

(Table values may vary.)

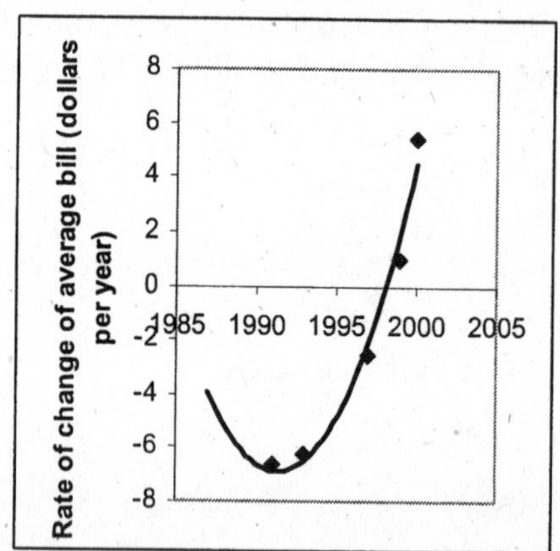

15. a. (Table values may vary slightly.)

Year	1985.25	1990	1995	1997	2000
Slope	8	46.8	79.2	55.2	21.4

b. Graph may vary, but its basic shape should be concave down with a maximum between 1993 and 1995.

17. a. The average rate of change during the year (found by estimating the slope of the secant line drawn from September to May) is approximately 14 members per month. (One possible answer.)

b,c. By estimating the slopes of tangent lines we obtain the following. (One possible answer.)

Month	Slope (members per month)
Sept	98
Nov	–9
Feb	30
Apr	11

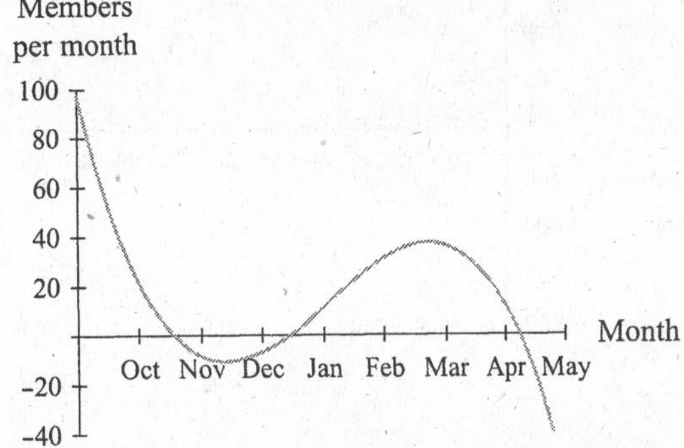

d. Membership was growing most rapidly around March. This point on the membership graph is an inflection point.

e. The average rate of change is not useful in sketching an instantaneous-rate-of-change graph.

19.

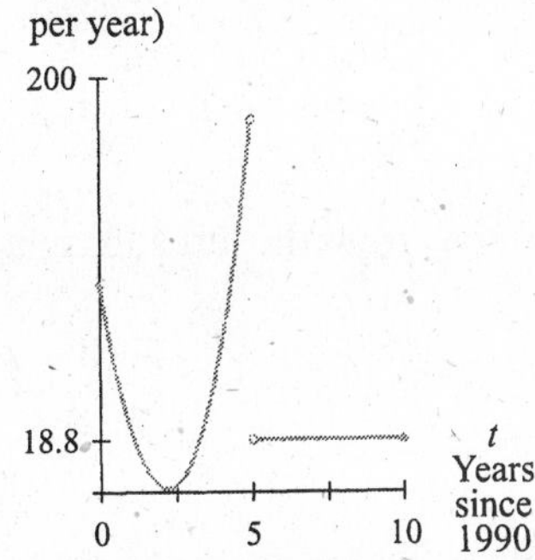

21. a. Slope $\approx \dfrac{\$18{,}000}{30 \text{ cars}} = \$600$ per car

Profit is increasing on average by approximately $600 per car.

b. By sketching tangent lines we obtain the following estimates: (Answers will vary depending on points picked and estimates of slopes.)

Number of cars	Slope (dollars per car)
20	0
40	160
60	750
80	10
100	–1200

c. Average monthly profit (dollars per car)

1200
900
600
300
-300
-600
-900
-1200

10 20 40 50 60 70 80 90 100 110 120 Cars sold

d. The average monthly profit is increasing most rapidly for approximately 60 cars sold and is decreasing most rapidly when approximately 100 cars are sold. The corresponding points on the graph are inflection points.

e. Average rates of change are not useful when graphing an instantaneous-rate-of-change graph.

23. The derivative does not exist at $x = 0$, $x = 3$, and $x = 4$ because the graph is not continuous at those input values.

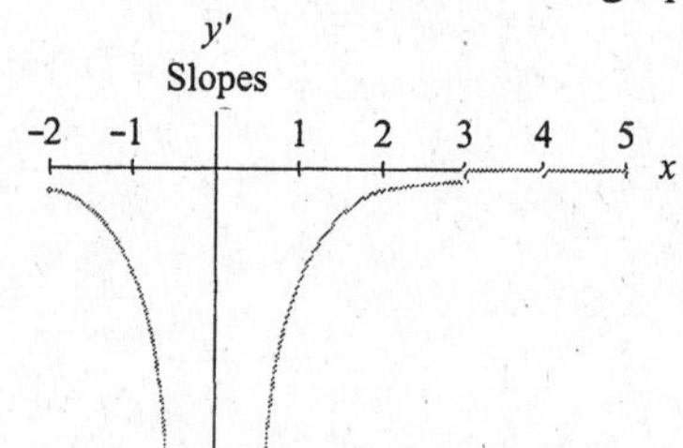

25. The derivative does not exist at $x = 2$ and $x = 3$ because the slopes from the left and right are different at those inputs.

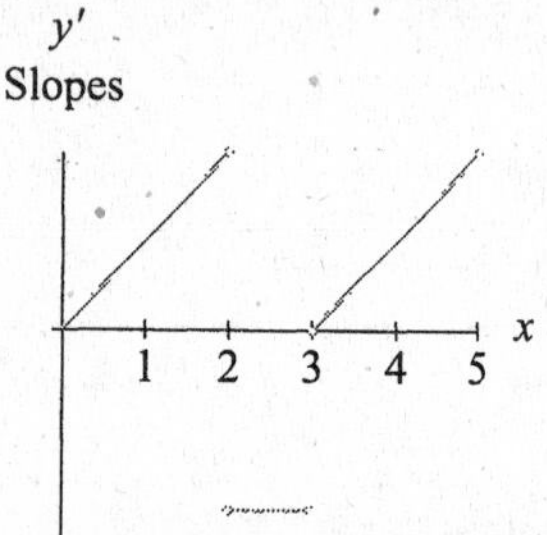

27. *One possible graph:*

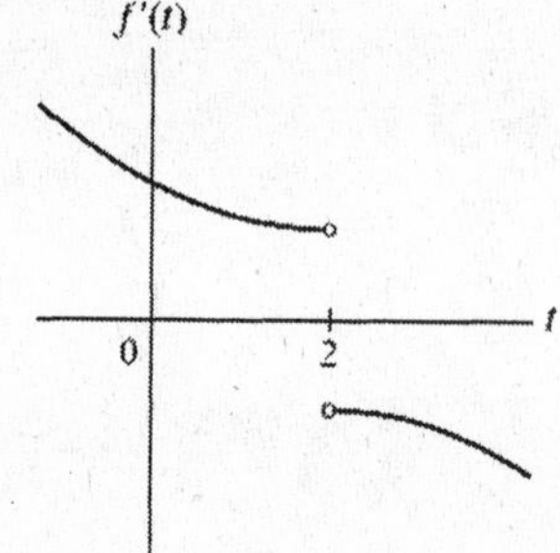

29. *One possible graph:*

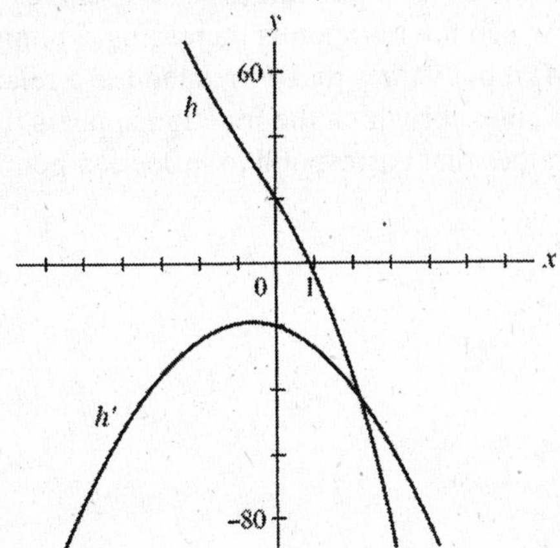

31. a.

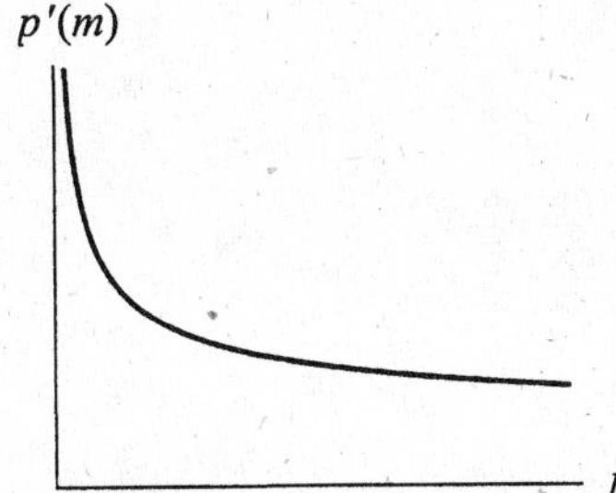

b. i. $p(m) = m + \sqrt{m}$

ii. $p(m+h) = m + h + \sqrt{m+h}$

iii. $\dfrac{p(m+h) - p(m)}{m+h-m} = \dfrac{h + \sqrt{m+h} - \sqrt{m}}{h}$

$$= \left(\frac{h + \sqrt{m+h} - \sqrt{m}}{h}\right)\frac{\left(\sqrt{m+h} + \sqrt{m}\right)}{\left(\sqrt{m+h} + \sqrt{m}\right)}$$

$$= \frac{h\left(\sqrt{m+h} + \sqrt{m}\right) + (m+h) - m}{h\left(\sqrt{m+h} + \sqrt{m}\right)} = \frac{h\left(\sqrt{m+h} + \sqrt{m}\right) + h}{h\left(\sqrt{m+h} + \sqrt{m}\right)}$$

iv. $\lim\limits_{h \to 0} \dfrac{h\left(\sqrt{m+h} + \sqrt{m}\right) + h}{h\left(\sqrt{m+h} + \sqrt{m}\right)} = \lim\limits_{h \to 0} \dfrac{\left(\sqrt{m+h} + \sqrt{m}\right) + 1}{\sqrt{m+h} + \sqrt{m}} = \dfrac{2\sqrt{m} + 1}{2\sqrt{m}} = 1 + \dfrac{1}{2\sqrt{m}}$

$\dfrac{dp}{dm} = 1 + \dfrac{1}{2\sqrt{m}}$. The graph of $\dfrac{dp}{dm}$ is the same as the one in part *a*.

33. *One possible answer:* When sketching a rate-of-change graph, it is important to identify the following features on the graph of the original function: 1) input values for which the derivative does not exist, 2) input intervals over which the function is increasing, 3) input intervals over which the function is decreasing, 4) input values that correspond to a relative maximum or minimum of the function, 5) input values for which the function appears to be increasing or decreasing most rapidly, 6) input values that correspond to inflection points where the function has zero slope.

Section 3.2 Simple Rate-of-Change Formulas

1.

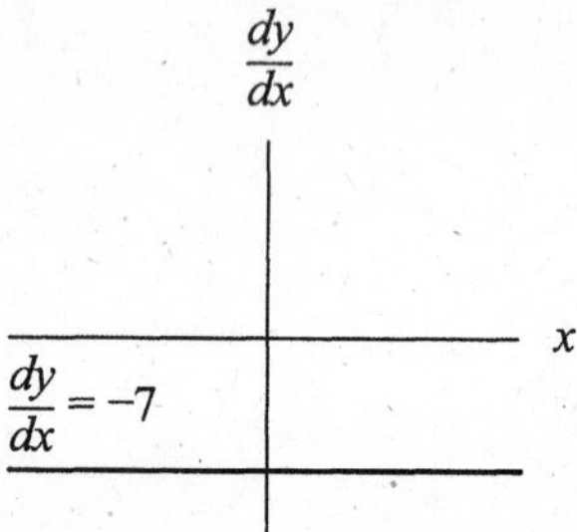

The function $y = 2 - 7x$ is a line with slope −7. Thus $\frac{dy}{dx} = -7$.

3.

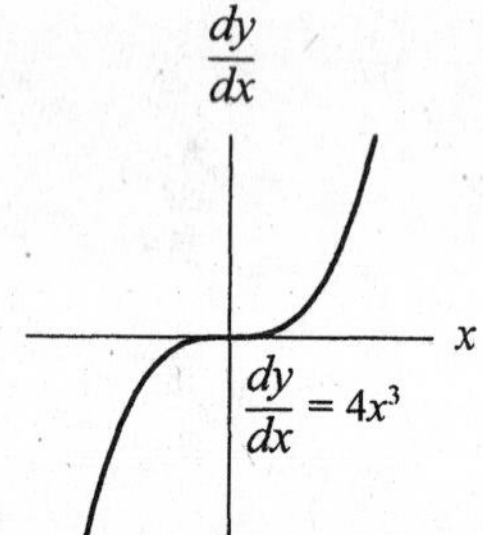

The slope formula is $\frac{dy}{dx} = 4x^{4-1} = 4x^3$.

5.

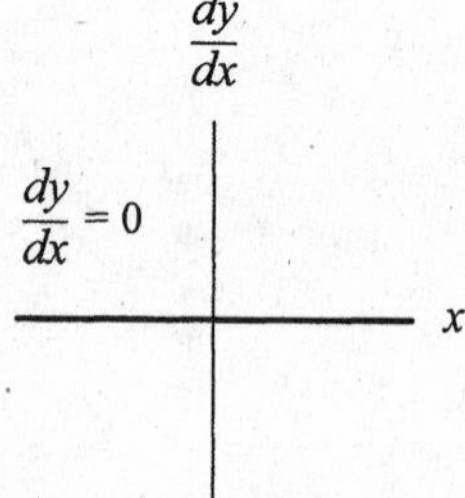

The slope of any horizontal line is 0.

7. The power rule applies: $f'(x) = 5x^{5-1} = 5x^4$

9. The power rule and the constant multiplier rule apply: $f'(x) = 3\left(3x^{3-1}\right) = 9x^2$

11. The linear function rule (or the power rule) apply: $f'(x) = -5$

13. The constant rule applies: $f'(x) = 0$

15. The linear function rule (or the power rule and the constant multiplier rule) apply: $\frac{dy}{dx} = 12$

17. The sum rule, the constant multiplier rule, the power rule, and the constant rule apply:

$$\frac{dy}{dx} = 5\left(3x^{3-1}\right) + 3\left(2x^{2-1}\right) - 2 = 15x^2 + 6x - 2$$

19. $f(x) = \frac{1}{x^3} = x^{-3} \rightarrow f'(x) = -3x^{-3-1} = -3x^{-4} = \frac{-3}{x^4}$

21. $f(x) = \frac{-9}{x^2} = -9x^{-2} \rightarrow f'(x) = -9\left(-2x^{-2-1}\right) = 18x^{-3} = \frac{18}{x^3}$

23. $f(x) = \frac{3x^2+1}{x} = \frac{3x^2}{x} + \frac{1}{x} = 3x + x^{-1} \rightarrow f'(x) = 3 - 1x^{-1-1} = 3 - x^{-2} = 3 - \frac{1}{x^2}$

25. $f(x) = \sqrt{x} = x^{\frac{1}{2}} \rightarrow f'(x) = \frac{1}{2}x^{\frac{1}{2}-1} = \frac{1}{2}x^{-\frac{1}{2}} = \frac{1}{2\sqrt{x}}$

27. a. $A'(t) = 0.1333$ dollars per year gives the rate of change in the average ATM transaction fee t years after 1990, $6 \le t \le 9$.

b. $A(10) = 0.1333(10) + 0.17 \approx \$1.50$

c. $A'(9) = \$0.1333$ per year

The transaction fee in 1999 was increasing by approximately $0.13 per year.

29. a. $T'(x) = -1.6t + 2$ °F per hour is the rate of change of the temperature t hours after noon

$T'(1.5) = -1.6(1.5) + 2 = -0.4$° F per hour

$T'(-5) = -1.6(-5) + 2 = 10$° F/hour

Steepness does not consider the sign of the slope, so we compare 0.4 and 10 and conclude that the graph is steeper at 7 a.m. than it is at 1:30 p.m.

b. $T'(-5) = -1.6(-5) + 2 = 10$°F per hour

c. $T'(0) = -1.6(0) + 2 = 2$°F per hour

d. $T'(4) = -1.6(4) + 2 = -4.4$°F per hour

Because this value is negative, the temperature is falling at a rate of 4.4°F per hour.

31. a,b. $B'(x) = 0.2685(3x^2) - 15.6(2x) + 94.684 + 0 = 0.8055x^2 - 31.2x + 94.684$ births per year is the rate of change in the number of live births to U.S. women aged 45 and over, where x is the number of years since 1950, $0 \le x \le 50$.

$B'(20) = 0.8055(20)^2 - 31.2(20) + 94.684 = -207.12$ births per year in 1970

$B'(45) = 0.8055(45)^2 - 31.2(45) + 94.684 = 321.82$ births per year in 1995

In 1970, the number of live births to U.S. women over 45 and over was falling by approximately 207 births per year. In 1995, the number of live births to this group was rising by approximately 322 births per year.

33. a. $m(w) = 6.930w + 682.188$ kilocalories per day is the metabolic rate of a typical 18- to 30-year old male who weighs w pounds, $88 \le w \le 200$.

b. $m'(w) = 6.930$ kilocalories per day per pound is the rate of change in the metabolic rate of a typical 18- to 30-year old male who weighs w pounds, $88 \le w \le 200$.

c. Regardless of the man's weight, if he gains weight, his metabolic rate will increase by 6.930 kilocalories per day per pound.

35. a. $R(x) = -3.68x^3 + 47.958x^2 - 80.759x + 166.98$ billion dollars is the revenue when \$$x$ billion is spent on advertising, $1.2 \le x \le 6.4$.

b. $R'(x) = -3.68(3x^2) + 47.958(2x) - 80.759(1) + 0 = -11.039x^2 + 95.916x - 80.759$ billion dollars per billion dollars (billion dollars of revenue per billion dollars of advertising) is the rate of change of revenue when \$$x$ billion is spent on advertising, $1.2 \le x \le 6.4$.

c. Revenue: $R(5) = -3.68(5)^3 + 47.958(5)^2 - 80.759(5) + 166.98 \approx \$502.166$ billion

Rate of Change of Revenue: $R'(5) = -11.039(5)^2 + 95.916(5) - 80.759 \approx \$122.837$ billion per billion dollar

d. $\dfrac{R'(5)}{R(5)}100\% \approx 24.46\%$ per billion dollars spent on advertising.

37. a. $P(x) = 175 - \left(0.015x^2 - 0.78x + 46 + \dfrac{49.6}{x}\right)$ dollars is the profit from the sale of one storm window when x windows are produced each hour, $x > 0$.

b. $P'(x) = 0 - \left[0.015(2x) - 0.78 + 0 + 49(-1x^{-2})\right] = -0.03x + 0.78 + \dfrac{49}{x^2}$

dollars per window gives the rate of change of the profit from the sale of one storm window when x windows are produced each hour, $x > 0$.

c. $P(80) = 175 - \left[0.015(80)^2 - 0.78(80) + 46 + \dfrac{49}{80}\right] \approx \$94.79$ profit

d. $P'(80) == -0.03(80) + 0.78 + \dfrac{49}{(80)^2} \approx -\$1.61$ per window produced

When 80 units are produced each hour, the profit from the sale of one window is decreasing by \$1.61 per unit produced. In other words, if production is increased from 80 to 81 windows per hour, then the average profit per window will decrease by approximately \$1.61. This indicates that if the company wishes to maximize the profit per window, they should not increase production above 80 units per hour.

39. *One possible answer:* The graph of a cubic function with a positive ax^3 term will increase, then decrease (or level off if the cubic function does not decrease at all), and finally increase again. The derivative graph for this cubic function will be a parabola with positive values at each end corresponding to the intervals of increase and either an interval of negative values (or the single slope value of 0 corresponding to the point that divides the two intervals of increase).

Section 3.3 Exponential and Logarithmic Rate-of-Change Formulas

1.

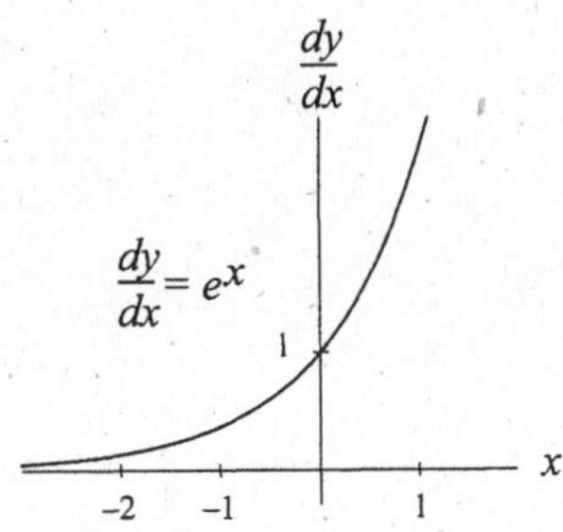

3.

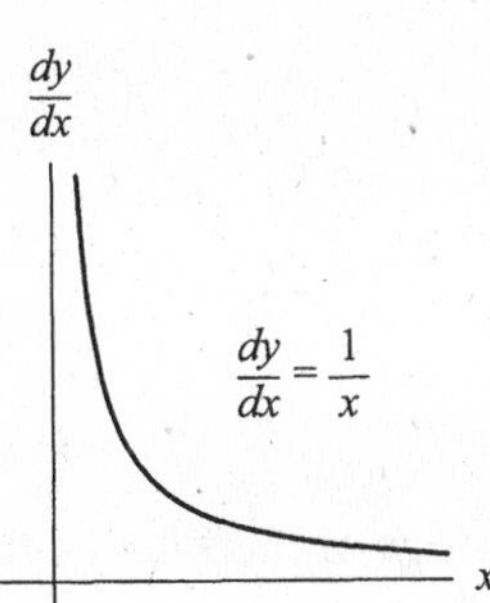

5.

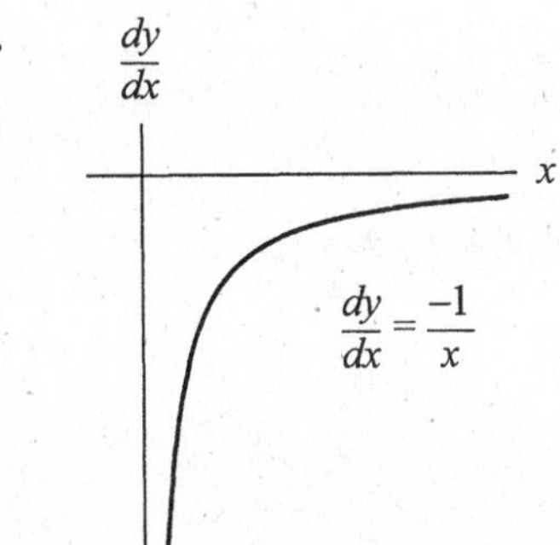

7. $h'(x) = 0 - 7e^x = -7e^x$

9. $\frac{dg}{dx} = (\ln 2.1)(2.1^x)$

11. $h'(x) = 12(\ln 1.6)(1.6^x)$

13. $f(x) = 10\left(\left(1+\dfrac{0.05}{4}\right)^4\right)^x$

$$f'(x) = 10\left(\ln\left(1+\tfrac{0.05}{4}\right)^4\right)\left(1+\tfrac{0.05}{4}\right)^{4x}$$

$$\approx 0.4969\left(1+\tfrac{0.05}{4}\right)^{4x}$$

$$\approx 0.4969(1.0509^x)$$

15. $j'(x) = 4.2(\ln 0.8)(0.8^x) + 0$

$$= 4.2(\ln 0.8)(0.8^x)$$

17. $j'(x) = 4 \cdot \dfrac{1}{x} = \dfrac{4}{x}$

19. $\dfrac{dy}{dx} = 0 - 7 \cdot \dfrac{1}{x} = \dfrac{-7}{x}$

21. a. $A(t) = \left(e^{0.043}\right)^t$

$$\frac{dA}{dt} = \ln(e^{0.043})e^{0.043t}$$

$$= 0.043(\ln e)e^{0.043t}$$

$= 0.043e^{0.043t}$ thousand dollars per year is the rate of change of the value of the investment after t years, $t > 0$.

b. $A(5) = e^{0.043(5)}$

≈ 1.23986 thousand dollars

$= \$1239.86$

c. $A'(5) = 0.043e^{0.043(5)}$

≈ 0.0533 thousand dollars per year

$\approx \$53$ per year

d. $\dfrac{A'(5)}{A(5)} \cdot 100 \approx 4.3$ % per year

23. a. $\dfrac{dA}{dr} = 1000\ln(e^{10})e^{10r} = 1000(10)e^{10r} = 10{,}000e^{10r}$

$= 10{,}000e^{10r}$ dollars per 100 percentage points when the interest rate is $100r$%

b. $A'(0.07) = 10{,}000e^{10(0.07)} \approx \$20{,}137.53$ per 100 percentage points. Working with interest can get interesting when taking derivatives. The increase of \$20,137.53 represents the rate of change when the interest rate (currently $r = 0.07$) increased to by 1 to 1.07. This means the interest rate is increased from 7% to 107%.

c. $\frac{dA}{dr} = 1000\ln(e^{0.1})e^{0.1r} = 1000(0.1)e^{0.1r} = 100e^{0.1r}$ dollars per percentage point when the interest rate is *r*%. The constant multiplier and exponent multiplier are both $\frac{1}{100}$ of what they were in the function in part *a*.

d. $A'(7) = 100e^{0.1(7)} \approx \$201.38$ per percentage point. This answer is $\frac{1}{100}$ of the answer to part *b*.

25. a. Solve $2.5 = 0.14\left(4.106^x\right)$

$$\frac{2.5}{0.14} = 4.106^x$$

$$\ln\left(\frac{2.5}{0.14}\right) = \ln\left(4.106^x\right)$$

$$\ln\left(\frac{2.5}{0.14}\right) = x\ln 4.106$$

$$\frac{\ln\left(\frac{2.5}{0.14}\right)}{\ln 4.106} = x$$

$$x \approx 2.041$$

b. $s'(x) = 0.14(\ln 4.106)(4.106^x)$ million iPods per year is the rate of change of iPod sales, where x is the number of years after the fiscal year that ended in September, 2002, $0 \le x \le 3$.

c. $s'(2.041) = 0.14(\ln 4.106)(4.106^{2.0407}) \approx 3.531$ million iPods per year; The iPod sales were increasing by approximately 3.531 million iPods per year at the time the 2.5 millionth iPod was sold.

27. a. 9 weeks: $w(7) = 11.3 + 7.37\ln 7 \approx 25.64$ grams; $w'(7) = \frac{7.37}{7} \approx 1.05$ grams per week

b. $\frac{w(9) - w(4)}{9 - 4} \approx \frac{5.977 \text{ grams}}{5 \text{ weeks}} \approx 1.195$ grams per week on average

c. The older the mouse is, the more slowly it gains weight because the rate-of-change formula, $w'(t) = \frac{7.37}{t}$, has the age of the mouse in the denominator.

29. a. $T'(d) = \frac{-9.9}{d}$ °F per day is the rate of change of the temperature at which mil must be stored in order to remain fresh, where d is the number of days that the milk must be stored, $d > 0$.

b. As the number of days that the milk must be stored increases, the temperature at which the milk must be stored decreases more slowly. Consequently, the rate of change of the temperature approaches zero.

31. a. $\dfrac{318.7-294.1}{1999-1997}=12.3$ index points per year; this is the average rate of change between 1997 and 1999.

b. $C(x)=-351.521+227.777\ln x$ points is the consumer price index for college tuition between 1990 and 2000, where x is the number of years since 1980.

c. $C'(18)=\dfrac{227.777}{18}\approx 12.7$ points per year

33. a. $R(x)=3.960(3.584^x)$ million dollars is the revenue realized by Apple from the sales of iPods for fiscal years ending between September 30, 2002, and September 30, 2006, where x is the number of fiscal years since September 30, 2000.

b. $R'(x)=3.960(\ln 3.584)(3.584^x)$ million dollars per year is the rate of change of the revenue realized by Apple from the sales of iPods for fiscal years ending between September 30, 2002, and September 30, 2006, where x is the number of fiscal years since September 30, 2000.

c. Revenue: $R(5)=3.960(3.584^5)\approx \$2340.44$

Rate of Change: $R'(5)=3.960(\ln 3.584)(3.584^5)=\$2987.53$ million per year

Percentage Rate of Change: $\dfrac{R'(5)}{R(5)}\cdot 100=127.65\,\%$ per year

The revenue realized by Apple for the fiscal year ending in September 30, 2005, from the sale of iPods was approximately $2340.44 million dollars. At that time, the revenue realized by Apple from the sale of iPods was increasing by approximately $2987.53 million per year. This rate of increase in the revenue realized by Apple from the sale of iPods was approximately 127.65 % per year.

35. a. $y=e^x \rightarrow \dfrac{dy}{dx}=(\ln e)(e^x)=(1)e^x=e^x$

b. $y=e^{kx}=\left(e^k\right)^x \rightarrow y'=\left(\ln e^k\right)\left(e^k\right)^x=k\left(e^k\right)^x=ke^{kx}$

Section 3.4 The Chain Rule

1. a. $f(x(2)) = f(6) = 140$

b. When $x = 6$, $\frac{df}{dx} = f'(6) = -27$

c. When $t = 2$, $\frac{dx}{dt} = x'(2) = 1.3$

d. When $t = 2$, $x = x(2) = 6$, so $\frac{df}{dt} = f'(6)x'(2) = (-27)(1.3) = -35.1$

3. Let t denote the number of days from today, w denote the weight in ounces, and v denote the value of the gold in dollars. We know that $\frac{dv}{dw} = \$395.70$ per troy ounce and $\frac{dw}{dt} = 0.2$ troy ounce per day. We seek the value of $\frac{dv}{dt}$:

$\frac{dv}{dt} = \frac{dv}{dw}\frac{dw}{dt} =$ (\$323.10 per troy ounce)(0.2 troy ounce per day)

= \$64.62 per day

The value of the investor's gold is increasing at a rate of \$64.62 per day.

5. a. $R(476) = \$10{,}000$ Canadian

On November 25, 2002, 476 units of the commodity were sold, producing revenue of \$10,000 Canadian.

b. $C(10{,}000) = \$633.47$

On November 25, 2002, 10,000 Canadian dollars were worth 633.47 U.S. dollars.

c. $\frac{dR}{dx} = \$2.6$ Canadian per unit; On November 25, 2002, the revenue was increasing by 2.6 Canadian dollars per unit sold.

d. $\frac{dC}{dr} = \$0.6335$ U.S. per Canadian dollar; On November 25, 2002, the exchange rate was \$0.6335 U.S. per Canadian dollar.

e. $\frac{dC}{dx} = \frac{dC}{dr}\frac{dR}{dx}$

= (\$0.6335 U.S. per Canadian dollar)(\$2.6 Canadian per unit)

= \$1.65 U.S. per unit

On November 25, 2002, revenue was increasing at rate of \$1.65 U.S. per unit sold.

7. a. In 2010, $t = 10$ and $p(10) = \frac{130}{1+12e^{-0.02(10)}} \approx 12.009$ thousand people

In 2010 the city had a population of approximately 12,000 people.

b. $g(p(10)) \approx g(12.009) \approx 2(12.009) - 0.001(12.009^3) \approx 22$ garbage trucks

In 2010 the city owned 22 garbage trucks.

c. Let $u = 1 + 12e^{-0.02t}$ and $v = -0.02t$. Then $p = 130u^{-1}$ and $u = 1 + 12e^{v}$.

$$p'(t) = \frac{dp}{dt} = \frac{dp}{du}\frac{du}{dt} = \frac{dp}{du}\frac{du}{dv}\frac{dv}{dt}$$

$$= \frac{d}{du}(130u^{-1})\frac{d}{dv}(1+12e^{v})\frac{d}{dt}(-0.02t) = (-130u^{-2})(12e^{v})(-0.02)$$

$$= \frac{31.2e^{v}}{u^2} = \frac{31.2e^{-0.02t}}{(1+12e^{-0.02t})^2} \text{ thousand people per year}$$

$$p'(10) = \frac{31.2e^{-0.02(10)}}{(1+12e^{-0.02(10)})^2} \approx 0.22 \text{ thousand people per year}$$

In 2010 the population was increasing at a rate of approximately 220 people per year.

d. $g'(p) = 2 - 0.001(3p^2)$

$= 2 - 0.003p^2$ trucks per thousand people

$g'(p(10)) \approx g'(12.009)$

$= 2 - 0.003(12.009^2) \approx 1.6$ trucks per thousand people

In 2010 when the population was approximately 12,000, the number of garbage trucks needed by the city was increasing by 1.6 trucks per thousand people.

e. $\frac{dg}{dt} = \frac{dg}{dp}\frac{dp}{dt} = g'(p(10))p'(10)$

$\approx$ (1.6 trucks per thousand people)(0.22 thousand people per year)

$\approx$ 0.34 trucks per year

In 2010 the number of trucks needed by the city was increasing at a rate of approximately 0.34 truck per year, or 1 truck every 3 years.

f. See interpretations in parts *a* through *e*.

9. $c(x(t)) = 3(4-6t)^2 - 2$

$$\frac{dc}{dt} = \frac{dc}{dx}\frac{dx}{dt} = (6x)(-6) = 6(4-6t)(-6) = -144 + 216t$$

11. $h(p(t)) = \dfrac{4}{1+3e^{-0.5t}}$

Let $u = 0.5t$. Then $p = 1 + 3e^{u}$.

$$\frac{dh}{dt} = \frac{dh}{dp}\frac{dp}{dt} = \frac{dh}{dp}\frac{dp}{du}\frac{du}{dt}$$

$$= \frac{d}{dp}(4p^{-1})\frac{d}{du}(1+3u^{4})\frac{d}{dt}(-0.5t)$$

$$= (-4p^{-2})(3e^{u})(-0.5)$$

$$= \frac{6e^{u}}{p^2}$$

$$= \frac{6e^{-0.5t}}{(1+3e^{-0.5t})^2}$$

13. $k(t(x)) = 4.3(\ln x)^3 - 2(\ln x)^2 + 4\ln x - 12$

$$\frac{dk}{dx} = 4.3\left[3(\ln x)^2\right]\frac{1}{x} - 2(2\ln x)\frac{1}{x} + 4\frac{1}{x} - 0$$

$$= \frac{12.9(\ln x)^2}{x} - \frac{4\ln x}{x} + \frac{4}{x}$$

15. $p(t(k)) = 7.9(1.046)^{14k^3-12k^2}$

$$\frac{dp}{dk} = 7.9(\ln 1.046)(1.046)^{14k^3-12k^2}(42k^2 - 24k)$$

17. Inside function: $u = 3.2x + 5.7$ Outside function: $f = u^5$

$$\frac{df}{dx} = \frac{df}{du}\frac{du}{dx} = \frac{d}{du}(u^5)\frac{d}{dx}(3.2x + 5.7)$$

$$= (5u^4)(3.2)$$

$$= 5(3.2x + 5.7)^4(3.2) = 16(3.2x + 5.7)^4$$

19. Inside function: $u = x - 1$ Outside function: $f = \frac{8}{u^3} = 8u^{-3}$

$$\frac{df}{dx} = \frac{df}{du}\frac{du}{dx}$$

$$= \frac{d}{du}(8u^{-3})\frac{d}{dx}(x - 1)$$

$$= (-24u^{-4})(1)$$

$$= \frac{-24}{u^4}$$

$$= \frac{-24}{(x-1)^4}$$

21. Inside function: $u = x^2 - 3x$ Outside function: $f = \sqrt{u} = u^{\frac{1}{2}}$

$$\frac{df}{dx} = \frac{df}{du}\frac{du}{dx} = \frac{d}{du}(u^{\frac{1}{2}})\frac{d}{dx}(x^2 - 3x)$$

$$= \left(\frac{1}{2}u^{-\frac{1}{2}}\right)(2x - 3)$$

$$= \frac{2x-3}{2\sqrt{u}} = \frac{2x-3}{2\sqrt{x^2 - 3x}}$$

23. Inside function: $u = 35x$ Outside function: $f = \ln u$

$$\frac{df}{dx} = \frac{df}{du}\frac{du}{dx}$$
$$= \frac{d}{du}(\ln u)\frac{d}{dx}(35x)$$
$$= \left(\frac{1}{u}\right)(35) = \frac{1}{35x}(35) = \frac{1}{x}$$

25. Inside function: $u = 16x^2 + 37x$ Outside function: $f = \ln u$

$$\frac{df}{dx} = \frac{df}{du}\frac{du}{dx} = \frac{d}{du}(\ln u)\frac{d}{dx}(16x^2 + 37x)$$
$$= \left(\frac{1}{u}\right)(32x + 37)$$
$$= \frac{1}{16x^2 + 37x}(32x + 37)$$
$$= \frac{32x + 37}{16x^2 + 37x}$$

27. Inside function: $u = 0.6x$ Outside function: $f = 72e^u$

$$\frac{df}{dx} = \frac{df}{du}\frac{du}{dx}$$
$$= \frac{d}{du}(72e^u)\frac{d}{dx}(0.6x)$$
$$= (72e^u)(0.6)$$
$$= (72e^{0.6x})(0.6)$$
$$= 43.2e^{0.6x}$$

29. Inside function: $u = 0.08x$ Outside function: $f = 1 + 58e^u$

$$\frac{df}{dx} = \frac{df}{du}\frac{du}{dx}$$
$$= \frac{d}{du}(1 + 58e^u)\frac{d}{dx}(0.08x)$$
$$= (58e^u)(0.08)$$
$$= (58e^{0.08x})(0.08) = 4.64e^{0.08x}$$

31. Inside function: $u = 1 + 18e^{0.6x}$ $\begin{cases} \text{inside: } w = 0.6x \\ \text{outside: } u = 1 + 18e^w \end{cases}$

Outside function: $f = \dfrac{12}{u} + 7.3 = 12u^{-1} + 7.3$

$$\frac{df}{dx}=\frac{df}{du}\frac{du}{dx}=\frac{df}{du}\frac{du}{dw}\frac{dw}{dx}$$
$$=\frac{d}{du}(12u^{-1}+7.3)\frac{d}{dw}(1+18e^{w})\frac{d}{dx}(0.6x)$$
$$=(-12u^{-2})(18e^{w})(0.6)$$
$$=-12(1+18e^{0.6x})^{-2}(18e^{0.6x})(0.6)$$
$$=\frac{-129.6e^{0.6x}}{(1+18e^{0.6x})^{2}}$$

33. Inside function: $u=\sqrt{x}-3x=x^{\frac{1}{2}}-3x$ Outside function: $f=u^3$

$$\frac{df}{dx}=\frac{df}{du}\frac{du}{dx}$$
$$=\frac{d}{du}(u^3)\frac{d}{dx}(x^{\frac{1}{2}}-3x)$$
$$=(3u^2)(\tfrac{1}{2}x^{\frac{-1}{2}}-3)$$
$$=3(x^{\frac{1}{2}}-3x)^2(\tfrac{1}{2}x^{\frac{-1}{2}}-3)=3(\sqrt{x}-3x)^2\left(\frac{1}{2\sqrt{x}}-3\right)$$

35. Inside function: $u=\ln x$ Outside function: $f=2^u$

$$\frac{df}{dx}=\frac{df}{du}\frac{du}{dx}$$
$$=\frac{d}{du}(2^u)\frac{d}{dx}(\ln x)$$
$$=(\ln 2)2^u\frac{1}{x}$$
$$=(\ln 2)2^{\ln x}\frac{1}{x}$$
$$=\frac{(\ln 2)2^{\ln x}}{x}$$

37. Inside function: $u=-Bx$ Outside function: $f=Ae^u$

$$\frac{df}{dx}=\frac{df}{du}\frac{du}{dx}$$
$$=\frac{d}{du}(Ae^u)\frac{d}{dx}(-Bx)$$
$$=Ae^u(-B)$$

39. a. $S(u(x))=0.75\sqrt{-2.3x^2+53.2x+249.8}+1.8$ millions of dollars is the predicted sales for a large firm, where x is the number of years in the future.

b. $\frac{d}{dx}S(u(x)) = \frac{dS}{du}\frac{du}{dx} = \frac{d}{du}(0.75u^{\frac{1}{2}} + 1.8)\frac{d}{dx}(-2.3x^2 + 53.2x + 249.8)$

$= \left[0.75\left(\frac{1}{2}u^{-\frac{1}{2}}\right)\right]\left[-2.3(2x) + 53.2 + 0\right] = \frac{0.375(-4.6x + 53.2)}{\sqrt{u}}$

$= \frac{-1.725x + 19.95}{\sqrt{-2.3x^2 + 53.2x + 249.8}}$ millions of dollars per year is the rate of change in predicted sales for a large firm, where x is the number of years in the future.

c. Answer will vary depending on the current year. If the year is 2007, then the answer for 2007 would be obtained by substituting $x = 0$ in the expression from part b, so the rate of change would be approximately 1.26 million dollars per year.

41. a. $R'(q) = 0.0314(0.62285)e^{0.62285q}$ million dollars/quarter is the rate of change in the revenue q quarters after the start of 1998.

b.

Quarter Ending	June 1998	June 1999	June 2000
$R(q)$ million dollars	3.0	4.2	18.8
$R'(q)$ million dollars per quarter	0.07	0.82	9.92
$\frac{R'(q)}{R(q)}\cdot 100$ % per quarter	2.3	19.5	52.7

43. a. We use the formula developed in Activity 38: $m'(x) = \frac{LABe^{-Bx}}{(1 + Ae^{-Bx})^2}$

with $L = 49$, $A = 36.0660$, and $B = 0.206743$.

$$m'(x) = \frac{LABe^{-Bx}}{(1 + Ae^{-Bx})^2} = \frac{(49)(36.0660)(0.206743)e^{-0.206743x}}{(1 + 36.0660e^{-0.206743x})^2} \approx \frac{365.363e^{-0.206743x}}{(1 + 36.0660e^{-0.206743x})^2}$$

states per year is the rate of change of states with national P.T.A. membership, where x is the number of years since 1895, $0 \le x \le 36$.

b. $m(7) = \frac{49}{1 + 36.0660e^{-0.206743(7)}} \approx 5$ states

c. 1915: $m'(20) \approx 2.4$ states per year
1927: $m'(32) \approx 0.4$ state per year

45. a. $f(x) = 13865.113(1.035^x)$ dollars is the projected tuition at a private 4-year college for the years between 2000 and 2010, x years after 2000.

b. Because $ae^{kx} = a(e^k)^x$, we solve for k in the equation $e^k = 1.035$. Therefore, $k = \ln 1.035 \approx 0.0344$ and $f(x) \approx 13865.113e^{0.0344x}$

c. For the first formula:

$$f'(x) = \frac{d}{dx}13865.113\left(1.035^x\right) = 13865.113(\ln 1.035)\left(1.035^x\right)$$

$$\approx 476.98(1.035^x) \text{ dollars per year}$$

For the second formula: Let $u = 0.0344x$. Then $f \approx 13865.113e^u$.

$$\frac{df}{dx} = \frac{df}{du}\frac{du}{dx}$$

$$= \frac{d}{du}(13865.113e^u)\frac{d}{dx}(0.0344x) = (13865.113e^u)(0.0344) \approx 476.98e^u$$

$$\approx 476.98e^{0.0344x} \text{ dollars per year}$$

d. Using the first model: $f'(8) \approx 476.98(1.035^8) \approx 632$ dollars per year

Using the second model: $f'(8) \approx 476.98e^{(0.0344 \cdot 8)} \approx 632$ dollars per year

The answers are identical.

47. a. The data are essentially concave up, suggesting a quadratic or exponential function. We choose a quadratic model because it fits the data slightly better than an exponential model. $t(x) = 7.763x^2 + 47.447x + 1945.893$ units is the average weekly production cost at a manufacturing company during the xth quarter after January 1, 2000, for the period from January 2000 through December 2003.

b. $C(t(x)) = 196.25 + 44.45\ln(7.763x^2 + 47.447x + 1945.893)$ dollars is the weekly production cost at a manufacturing company during the xth quarter after January 1, 2000, for the period from January 2000 through December 2003.

c.

Jan–Mar 2004:	$C(t(17)) \approx \$574.81$ per week
Apr–June 2004:	$C(t(18)) \approx \$577.56$ per week
July–Sept 2004:	$C(t(19)) \approx \$580.27$ per week
Oct–Dec 2004:	$C(t(20)) \approx \$582.94$ per week

d.

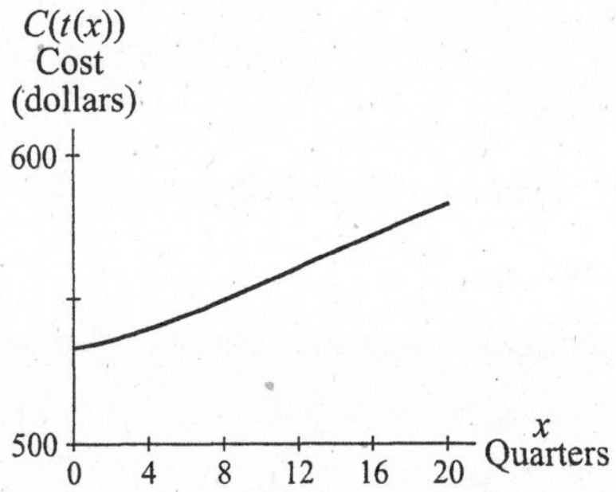

C(t(x))
Cost
(dollars)
600
500
0 4 8 12 16 20
x
Quarters

According to the graph, the cost appears never to decrease between January of 2000 and January of 2005.

e. $\frac{d}{dx}C(t(x)) = \frac{dC}{dt}\frac{dt}{dx}$

$$= \frac{d}{dt}(196.25 + 44.25\ln t)\frac{d}{dx}(7.763x^2 + 47.447x + 1945.893)$$

$$= 44.25\left(\frac{1}{t}\right)(15.525x + 47.447 + 0)$$

$= \frac{44.25(15.525x + 47.447)}{7.763x^2 + 47.447x + 1945.893}$ dollars per quarter is the rate of change of weekly manufacturing costs at a manufacturing company during the xth quarter after January 1, 2000, for the period from January 2000 through December 2003.

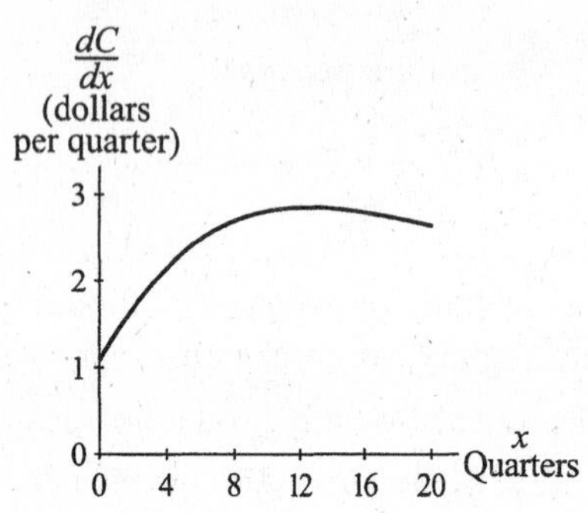

According to the graph, the rate of change is always positive. No, the cost will never decrease.

49. *One possible answer:* Composite functions are formed by making the output of one function (the inside) the input of another function (the outside). It is imperative that the output of the inside and the input of the outside agree in the quantity that they measure as well as in the units of measurement.

Section 3.5 The Product Rule

1. $h'(2) = f'(2)g(2) + f(2)g'(2) = -1.5(4) + 6(3) = 12$

3. a. i. In 2007 there were 75,000 households in the city.

ii. In 2007 the number of households was declining at a rate of 1200 per year.

iii. In 2007, 52% of households owned a computer.

iv. In 2007 the percentage of households with a computer was increasing by 5 percentage points per year.

b. Input: the number of years since 1995
Output: the number of households with computers

c. $N(2) = h(2)c(2) = (75{,}000)(0.52) = 39{,}000$ households with computers.
$N'(2) = h'(2)c(2) + h(2)c'(2) = (-1200)(0.52) + (75{,}000)(0.05)$
$= 3126$ households per year
In 2007 there were 39,000 households with computers, and that number was increasing at a rate of 3126 households per year.

5. a. i. $S(10) = 15 + \dfrac{2.6}{10+1} \approx \$15.24$

$S'(x) = \dfrac{d}{dx}[15 + 2.6(x+1)^{-1}] = 0 + 2.6(-1)(x+1)^{-2}\dfrac{d}{dx}(x+1)$

$= \dfrac{-2.6}{(x+1)^2}$ dollars per week

$S'(10) = \dfrac{-2.6}{(10+1)^2} \approx -\$0.02$ per week

After 10 weeks, 1 share is worth \$15.24, and the value is declining by \$0.02 per week.

ii. $N(10) = 100 + 0.25(10^2) = 125$ shares

$N'(x) = \dfrac{d}{dx}(100 + 0.25x^2) = 0.5x$ shares per week after x weeks

$N'(10) = 0.5(10) = 5$ shares per week

After 10 weeks, the investor owns 125 shares and is buying 5 shares per week.

iii. $V(10) = S(10)N(10) \approx (15.24)(125) \approx \$1904.55$

$V'(10) \approx S'(10)N(10) + S(10)N'(10)$

$\approx (-0.02)(125) + (15.24)(5) \approx \$73.50$ per week

After 10 weeks, the investor's stock is worth approximately \$1905, and the value is increasing at a rate of \$73.50 per week.

b. $V'(x) = S'(x)N(x) + S(x)N'(x)$

$= \left(\dfrac{-2.6}{(x+1)^2}\right)(100 + 0.25x^2) + \left(15 + \dfrac{2.6}{x+1}\right)(0.5x)$

$= -\dfrac{0.65x^2 + 260}{(x+1)^2} + \dfrac{1.3x}{x+1} + 7.5x$ dollars per week after x weeks

7. Let $A(t)$ be the number of acres of corn and let $B(t)$ be the number of bushels of corn per acre, where t is the number of years from the current year. The total number of bushels of corn produced is given by $C(t) = A(t)B(t)$, where $A(0) = 500$ acres, $A'(0) = 50$ acres per year, $B(0) = 130$ bushels per acre, and $B'(0) = 5$ bushels per acre per year. The rate of change is:

$C'(0) = A'(0)B(0) + A(0)B'(0)$

$= (50 \text{ acres/year})(130 \text{ bushels/acre}) + (500 \text{ acres})(5 \text{ bushels/acre/year})$

$= 9000$ bushels per year

9. a. $(17{,}000)(0.48) = 8160$ voters

b. $(8160)(0.57) \approx 4651$ votes for candidate A

c. Let $v(t)$ be the proportion of registered voters who plan to vote, and let $a(t)$ be the proportion who support candidate A, where t is the number of weeks from today and both quantities are expressed as decimals. Then the number of votes for candidate A is given by

$N(t) = 17{,}000v(t)a(t)$, where $v(0) = 0.48$, $a(0) = 0.57$, $v'(0) = 0.07$ and $a'(0) = -0.03$.

The rate of change of N is:

$$N'(t) = \frac{d}{dt}[17{,}000v(t)a(t)] = 17{,}000\frac{d}{dt}[v(t)a(t)] = 17{,}000[v'(t)a(t) + v(t)a'(t)]$$

When $t = 0$,

$$N'(0) = 17{,}000[(0.07)(0.57) + (0.48)(-0.03)] \approx 434 \text{ votes for candidate A per week}$$

11. $$f'(x) = \left[\frac{d}{dx}(\ln x)\right]e^x + (\ln x)\frac{d}{dx}e^x$$
$$= \frac{1}{x}e^x + (\ln x)e^x$$
$$= \frac{e^x}{x} + (\ln x)e^x$$

13. $$f'(x) = \left[\frac{d}{dx}(3x^2 + 15x + 7)\right](32x^3 + 49) + (3x^2 + 15x + 7)\frac{d}{dx}(32x^3 + 49)$$
$$= (6x + 15)(32x^3 + 49) + (3x^2 + 15x + 7)(96x^2)$$
$$= 480x^4 + 1920x^3 + 672x^2 + 294x + 735$$

15. $$f'(x) = \left[\frac{d}{dx}(12.8x^2 + 3.7x + 1.2)\right][29(1.7^x)] + (12.8x^2 + 3.7x + 1.2)\frac{d}{dx}[29(1.7^x)]$$
$$= (25.6x + 3.7)[29(1.7^x)] + (12.8x^2 + 3.7x + 1.2)[29(\ln 1.7)(1.7^x)]$$

17. Note that $f(x) = g(x)h(x)$, where $g(x) = (5.7x^2 + 3.5x + 2.9)^3$ and $h(x) = (3.8x^2 + 5.2x + 7)^{-2}$.

$$g'(x) = 3(5.7x^2 + 3.5x + 2.9)^2\frac{d}{dx}(5.7x^2 + 3.5x + 2.9)$$
$$= 3(5.7x^2 + 3.5x + 2.9)^2(11.4x + 3.5)$$
$$h'(x) = -2(3.8x^2 + 5.2x + 7)^{-3}\frac{d}{dx}(3.8x^2 + 5.2x + 7)$$
$$= -2(3.8x^2 + 5.2x + 7)^{-3}(7.6x + 5.2)$$
$$f'(x) = g'(x)h(x) + g(x)h'(x)$$
$$= [3(5.7x^2 + 3.5x + 2.9)^2(11.4x + 3.5)](3.8x^2 + 5.2x + 7)^{-2}$$
$$+ (5.7x^2 + 3.5x + 2.9)^3[-2(3.8x^2 + 5.2x + 7)^{-3}(7.6x + 5.2)]$$

19. $f'(x) = \frac{d}{dx}[12.6(4.8^x)(x^{-2})]$

$= 12.6\frac{d}{dx}[(4.8^x)(x^{-2})]$

$= 12.6\left[\left(\frac{d}{dx}(4.8^x)\right)x^{-2} + (4.8^x)\frac{d}{dx}x^{-2}\right]$

$= 12.6\left[(\ln 4.8)(4.8^x)(x^{-2}) + (4.8^x)(-2x^{-3})\right]$

$= 12.6(4.8^x)\left(\frac{\ln 4.8}{x^2} - \frac{2}{x^3}\right) = \frac{12.6(4.8^x)}{x^3}[x(\ln 4.8) - 2]$

21. Note that $f(x) = g(x)h(x)$, where $g(x) = 79x$,

$h(x) = 198(1 + 7.68e^{-0.85x})^{-1} + 15$, and $g'(x) = 79$

Using the formula for the derivative of a logistic function, we know that

$h'(x) = \frac{LABe^{-Bx}}{(1 + Ae^{-Bx})^2}$ where $L = 198$, $A = 7.68$, and $B = 0.85$

Thus $h'(x) = \frac{198(7.68)(0.85)e^{-0.85x}}{(1 + 7.68e^{-0.85x})^2}$.

$f'(x) = g'(x)h(x) + g(x)h'(x)$

$= 79\left(\frac{198}{1 + 7.68e^{-0.85x}} + 15\right) + (79x)\left(\frac{198(7.68)(0.85)e^{-0.85x}}{(1 + 7.68e^{-0.85x})^2}\right)$

$= \frac{15{,}642}{1 + 7.68e^{-0.85x}} + 1{,}185 + \frac{102{,}110.976xe^{-0.85x}}{(1 + 7.68e^{-0.85x})^2}$

23. Note that $f(x) = g(x)h(x)$, where $g(x) = 430(0.62^x)$ and $h(x) = \left[6.42 + 3.3(1.46^x)\right]^{-1}$.

$g'(x) = 430(\ln 0.62)(0.62^x)$

$h'(x) = -\left[6.42 + 3.3(1.46^x)\right]^{-2} 3.3(\ln 1.46)(1.46^x) = \frac{-3.3(\ln 1.46)(1.46^x)}{\left[6.42 + 3.3(1.46^x)\right]^2}$

$f'(x) = g'(x)h(x) + g(x)h'(x)$

$= 430(\ln 0.62)(0.62^x)\frac{1}{6.42 + 3.3(1.46^x)} + 430(0.62^x)\frac{-3.3(\ln 1.46)(1.46^x)}{\left[6.42 + 3.3(1.46^x)\right]^2}$

25. $f'(x) = \left(\frac{d}{dx}4x\right)\sqrt{3x+2} + 4x\frac{d}{dx}(3x+2)^{1/2} + \frac{d}{dx}93$

$= 4\sqrt{3x+2} + 4x\left[\frac{1}{2}(3x+2)^{-1/2}(3)\right] + 0$

$= 4\sqrt{3x+2} + \frac{6x}{\sqrt{3x+2}}$

27. Note that $f(x) = g(x)h(x)$ where $g(x) = x$ and h is a logistic function with $L = 14$, $A = 12.6$, and $B = 0.73$ and derivative of the form $h'(x) = \frac{LABe^{-Bx}}{(1+Ae^{-Bx})^2}$:

$$h'(x) = \frac{128.772e^{-0.73x}}{\left(1+12.6e^{-0.73x}\right)^2}$$

$$f'(x) = g'(x)h(x) + g(x)h'(x) = (1)\frac{14}{1+12.6e^{-0.73x}} + (x)\frac{128.772e^{-0.73x}}{\left(1+12.6e^{-0.73x}\right)^2}$$

$$= \frac{14}{1+12.6e^{-0.73x}} + \frac{128.772xe^{-0.73x}}{\left(1+12.6e^{-0.73x}\right)^2}$$

29. a. $E(x) = \frac{0.73(1.2912^x)+8}{100}(-0.026x^2 - 3.842x + 538.868)$ women received epidural pain relief during childbirth at an Arizona hospital between 1981 and 1997, x years after 1980.

$E'(x) = \frac{0.73(1.2912^x)+8}{100}(-0.052x - 3.842) +$
$\frac{0.73(\ln 1.2912)(1.2912^x)}{100}(-0.026x^2 - 3.842x + 538.868)$ women per year is the rate of change in the number of women who received epidural pain relief during childbirth at an Arizona hospital between 1981 and 1997, x years after 1980.

b. Increasing by $p'(17) \approx 14.4$ percentage points per year

c. Decreasing by approximately 5 births per year ($b'(17) \approx -4.7$)

d. Increasing by $E'(17) \approx 64$ women per year.

e. Profit = $\$57 \cdot E(17) \approx \$17{,}043$ (using a value of 299 for the number of births) or \$17,071 (using an unrounded number of births)

31. a. Multiply the number of CDs, $6250(0.9286^x)$, by the price, x.

$R(x) \approx 6250x(0.9286^x)$ dollars is the monthly revenue from CD sales when the price of a CD is x dollars.

b. The profit for each CD is $x - 7.5$ dollars. Multiply the number of CDs, $6250(0.9286^x)$, by the per-CD profit, $x - 7.5$.

$P(x) = 6250(0.9286)^x(x - 7.5)$ dollars is the monthly profit from CD sales when the price of a CD is x dollars.

c. $R'(x) = 6250\frac{d}{dx}[x(0.9286)^x]$

$= 6250\left[\left(\frac{d}{dx}x\right)(0.9286)^x + x\frac{d}{dx}(0.9286)^x\right]$

$= 6250\left[(1)(0.9286^x) + x(\ln 0.9286)(0.9286^x)\right]$

$= 6250(0.9286^x)(1 + x\ln 0.9286)$ dollars of revenue per dollar of price is the rate of change in the monthly revenue from CD sales when the price of a CD is x dollars.

$P'(x) = 6250\frac{d}{dx}[(0.9286)^x(x - 7.5)]$

$= 6250\left[\left(\frac{d}{dx}0.9286^x\right)(x - 7.5) + (0.9286^x)\frac{d}{dx}(x - 7.5)\right]$

$= 6250[(\ln 0.9286)(0.9286^x)(x - 7.5) + (0.9286^x)(1)]$

$= 6250(0.9286)^x[1 + (x - 7.5)(\ln 0.9286)]$ dollars of profit per dollar of price is the rate of change in the monthly revenue from CD sales when the price of a CD is x dollars.

d. Evaluate the expressions for $R'(x)$ and $P'(x)$ to complete the table.

Price	Rate of change of revenue (dollars of revenue per dollar of price)	Rate of change of profit (dollars of profit per dollar of price)
\$13	88.27	1413.83
\$14	–82.15	1148.76
\$20	–684.05	105.17
\$21	–732.93	–0.06
\$22	–771.34	–90.79

e. Because $R'(13) > 0$ and $R'(14) < 0$, the highest revenue will be achieved with a price between \$13 and \$14.

f. Because $P'(x) > 0$ for $13 \le x \le 20, P'(21) \approx 0$, and $P'(21) < 0$, the price corresponding to the maximum profit is approximately \$21.

33. a. $C(x) = 71.459(1.050^x)$ dollars is the cost to produce x units in an hour, $10 \le x \le 90$.

b. $C'(x) = 71.459(\ln 1.050)(1.050^x)$ dollars per unit produced is the rate of change of the cost to produce x units in an hour, $10 \le x \le 90$.

c. $A(x) = \frac{C(x)}{x} = 71.459(1.050^x)(x^{-1})$ dollars per unit is the average cost to produce one unit when x units are produced each hour, $10 \le x \le 90$.

d. $A'(x) = \frac{d}{dx}[71.459(1.050^x)(x^{-1})]$

$$= 71.459\frac{d}{dx}[(1.050^x)(x^{-1})]$$

$$= 71.459\left[\left(\frac{d}{dx}1.050^x\right)(x^{-1}) + (1.050^x)\frac{d}{dx}(x^{-1})\right]$$

$$= 71.459\left[(\ln 1.050)(1.050^x)(x^{-1}) + (1.050^x)(-x^{-2})\right]$$

$= 71.459(1.050^x)\left(\frac{\ln 1.050}{x} - \frac{1}{x^2}\right)$ dollars per unit per hourly unit produced is the rate of change in the average cost to produce one unit when x units are produced, $10 \le x \le 90$.

e. 15 units: $A'(15) \approx -\$0.18$ per unit per hourly unit produced
35 units: $A'(35) \approx \$0.23$ per unit per hourly unit produced
85 units: $A'(85) \approx \$1.97$ per unit per hourly unit produced

f.

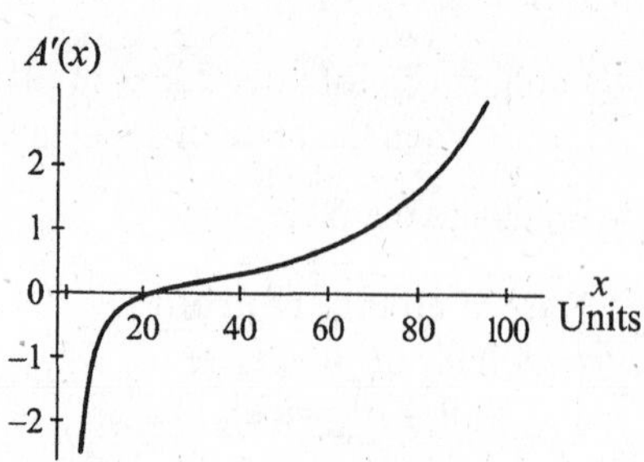

Yes, the average cost is decreasing when $A'(x) < 0$, for an hourly production of between 0 and 20 units.

g. The graph of A' crosses the x-axis near $x = 20.5$. Practically speaking, the average cost is decreasing at 20 units and increasing at 21 units, so a production level of 21 units is the one at which average cost first increases.

35. a. $t(x) = -0.025x^2 + 2.099x + 58.821$ million gives the number of households with TVs in the U.S. between 1970 and 2002, x years after 1970.

b. $v(x) = \frac{0.815}{1 + 31{,}239.748e^{-0.618x}}$ percent (expressed as a decimal) gives the percentage of U.S. households with TVs that also have VCRs between 1970 and 2002, x years after 1970.

c. $n(x) = t(x)v(x) = (-0.025x^2 + 2.099x + 58.821)\frac{0.815}{1 + 31{,}239.748e^{-0.618x}}$ million households owned TVs with VCRs x years after 1970, $0 \le x \le 32$.

d. $t'(x)=\frac{d}{dx}(-0.025x^2+2.099x+58.821)$

$=-0.050x+2.099$ million households per year x years after 1970

Note that v is a logistic function with $L=0.815$, $A=31{,}239.748$, and $B=0.618$ and derivative of the form $v'(x)=\frac{LABe^{-Bx}}{(1+Ae^{-Bx})^2}$:

$$v'(x)=\frac{(0.815)(31{,}239.748)(0.618)e^{-0.618x}}{(1+31{,}239.748e^{-0.618x})^2}$$

$$\approx\frac{15{,}658.143e^{-0.618x}}{(1+31{,}239.748e^{-0.618x})^2}$$

$$n'(t)=t'(x)v(x)+t(x)v'(x)$$

$$=(-0.050x+2.099)\left(\frac{0.815}{1+31{,}239.748e^{-0.618x}}\right)$$

$$+\left(-0.025x^2+2.099x+58.821\right)\left(\frac{15{,}658.143e^{-0.618x}}{(1+31{,}239.748e^{-0.618x})^2}\right)$$

million households per year is the rate of change in the number of households with TVs and VCRs x years after 1970, $0\le x\le 32$.

e. 1980: $n'(10)\approx 0.6$ million households per year

1985: $n'(15)\approx 8.4$ million households per year

1990: $n'(20)\approx 5.5$ million households per year

37. a. $E(x)=-151.516x^3+2060.988x^2-8819.062x+195{,}291.201$ students is the enrollment in ninth through twelfth grades in South Carolina between 1980-81 and 1989-90, where x is the number of years since the 1980-91 school year.

$D(x)=-14.271x^3+213.882x^2-1393.655x+11{,}697.292$ students dropping out from ninth through twelfth grades in South Carolina between 1980-81 and 1989-90, where x is the number of years since the 1980-91 school year.

b. $P(x)=\frac{D(x)}{E(x)}\cdot 100\,\%$ is the percent of students dropping out from ninth through twelfth grades in South Carolina between 1980-81 and 1989-90, where x is the number of years since the 1980–81 school year.

c. $D'(x) = -42.813x^2 + 427.763x - 1393.655$; $E'(x) = -454.548x^2 + 4121.976x - 8819.062$

Write $P(x) = 100D(x)[E(x)]^{-1}$.

$$P'(x) = 100\frac{d}{dx}\left(D(x)[E(x)]^{-1}\right)$$

$$= 100\left(\frac{d}{dx}D(x)\right)[E(x)]^{-1} + 100D(x)\frac{d}{dx}[E(x)]^{-1}$$

$$= 100D'(x)[E(x)]^{-1} + 100D(x)(-1)[E(x)]^{-2}E'(x)$$

$$= \frac{100D'(x)}{E(x)} - \frac{100D(x)\cdot E'(x)}{[E(x)]^2}$$

$$= \frac{100(-42.813x^2 + 427.763x - 1393.655)}{-151.516x^3 + 2060.988x^2 - 8819.062x + 195{,}291.201} -$$

$$\frac{100(-14.271x^3 + 213.882x^2 - 1393.655x + 11{,}697.292)(-454.548x^2 + 4121.976x - 8819.062)}{(-151.516x^3 + 2060.988x^2 - 8819.062x + 195{,}291.201)^2}$$

percentage points per year is the rate of change in the percent of students dropping out from ninth through twelfth grades in South Carolina between 1980-81 and 1989-90, where x is the number of years since the 1980–81 school year.

d.

X	$P'(x)$ (percentage point per year)	X	$P'(x)$ (percentage point per year)
0	–0.44	5	–0.19
1	–0.38	6	–0.19
2	–0.32	7	–0.22
3	–0.26	8	–0.29
4	–0.21	9	–0.41

In the 1980–81 school year, the rate of change was most negative with a value of –0.44 percentage point per year. This is the most rapid decline during this time period. The rate of change was least negative in the 1985–86 school year with a value of –0.187 percentage point per year.

e. Negative rates of change indicate that the number of dropouts in South Carolina was falling during the 1980s. This means that more students were staying in school.

39. *One possible answer:* For a function constructed by multiplying two other functions, the input is the same as the common input of the two other functions.

Section 3.6 Limiting Behavior Revisited: L'Hôpital's Rule

1. $\lim_{x\to 2} 2x^3 - 3^x = 2(2^3) - 3^2 = 7$ using direct substitution

3. $\lim_{x\to 0} e^x - \ln(x+1) = e^0 - \ln(1) = 1 - 0 = 1$ using direct substitution

5. $\lim_{x\to 0^+} \ln x = -\infty$ so $\lim_{x\to 0^+} \dfrac{1}{\ln x} = \dfrac{1}{-\infty} = 0$

7. $\lim_{n\to 1} \ln n = 0$ and $\lim_{n\to 1} n - 1 = 0$. Therefore, $\lim_{n\to 1} \dfrac{\ln n}{n-1}$ is of the indeterminate form $\dfrac{0}{0}$.

Applying L'Hopital's Rule: $\lim_{n\to 1} \dfrac{\ln n}{n-1} = \lim_{n\to 1} \dfrac{1/n}{1} = \lim_{n\to 1} 1/1 = 1$

9. $\lim_{x\to 1} x^4 - 1 = 0$ and $\lim_{x\to 1} x^3 - 1 = 0$. Therefore, $\lim_{x\to 1} \dfrac{x^4 - 1}{x^3 - 1}$ is of the indeterminate form $\dfrac{0}{0}$.

Applying L'Hopital's Rule: $\lim_{x\to 1} \dfrac{x^4 - 1}{x^3 - 1} = \lim_{x\to 1} \dfrac{4x^3}{3x^2} = \lim_{x\to 1} \dfrac{4}{3} x = \dfrac{4}{3}$.

11. $\lim_{t\to 0^+} \sqrt{t} \ln t$ is of the indeterminate form $0 \cdot -\infty$. Applying L'Hopital's Rule:

$$\lim_{t\to 0^+} \sqrt{t} \ln t = \lim_{t\to 0^+} \frac{\ln t}{t^{-1/2}} = \lim_{t\to 0^+} \frac{1/t}{-\frac{1}{2} t^{-3/2}} = \lim_{t\to 0^+} \frac{-2t^{3/2}}{t} = \lim_{t\to 0^+} -2t^{1/2} = -2(0) = 0\,.$$

13. $\lim_{x\to \infty} x^2 e^{-x}$ is of the indeterminate form $\infty \cdot 0$. Applying L'Hopital's Rule:

$$\lim_{x\to \infty} x^2 e^{-x} = \lim_{x\to \infty} \frac{x^2}{e^x} = \lim_{x\to \infty} \frac{2x}{e^x} = \lim_{x\to \infty} \frac{2}{e^x} = 0\ .$$

15. $\lim_{x\to 2} \dfrac{3x - 6}{x+2} = \dfrac{3(2) - 6}{2+2} = 0$ using direct substitution

17. $\lim_{x\to 5} \sqrt{x-1} - 2 = 0$ and $\lim_{x\to 5} x^2 - 25 = 0$. Therefore, $\lim_{x\to 5} \dfrac{\sqrt{x-1} - 2}{x^2 - 25}$ is of the indeterminate form $\dfrac{0}{0}$. Applying L'Hopital's Rule: $\lim_{x\to 5} \dfrac{\frac{1}{2}(x-1)^{-1/2}}{2x} = \lim_{x\to 5} \dfrac{1}{4x\sqrt{x-1}} = \dfrac{1}{4(5)\sqrt{5-1}} = \dfrac{1}{40}$.

19. $\lim_{x\to 2} 2x^2 - 5x + 2 = 0$ and $\lim_{x\to 2} 5x^2 - 7x - 6 = 0$. Therefore, $\lim_{x\to 2} \dfrac{2x^2 - 5x + 2}{5x^2 - 7x - 6}$ is of the indeterminate form $\dfrac{0}{0}$. Applying L'Hopital's Rule: $\lim_{x\to 2} \dfrac{2x^2 - 5x + 2}{5x^2 - 7x - 6} = \lim_{x\to 2} \dfrac{4x - 5}{10x - 7} = \dfrac{3}{13}$

21. $\lim_{x\to 0} (3x)\left(\dfrac{2}{e^x}\right) = \lim_{x\to 0} (3x) \cdot \lim_{x\to 0} \left(\dfrac{2}{e^x}\right) = 3(0) \cdot \left(\dfrac{2}{e^0}\right) = 0$

23. $\lim\limits_{x\to\infty}(.6^x)=0$ and $\lim\limits_{x\to\infty}(\ln x)=\infty$ so $\lim\limits_{x\to\infty}(.6^x)(\ln x)$ is of the indeterminate form $0\cdot\infty$.

Applying L'Hopital's Rule: $\lim\limits_{x\to\infty}(.6^x)(\ln x)=\lim\limits_{x\to\infty}\dfrac{\ln x}{.6^{-x}}=\lim\limits_{x\to\infty}\dfrac{1/x}{(\ln .6)(.6^{-x})(-1)}=\lim\limits_{x\to\infty}\dfrac{-(.6^x)}{x(\ln .6)}=0.$

25. $\lim\limits_{x\to\infty}3x^2+2x+4=\infty$ and $\lim\limits_{x\to\infty}5x^2+x+1=\infty$. Therefore, $\lim\limits_{x\to\infty}\dfrac{3x^2+2x+4}{5x^2+x+1}$ is of the indeterminate form $\dfrac{\infty}{\infty}$. Applying L'Hopital's Rule:

$$\lim_{x\to\infty}\frac{3x^2+2x+4}{5x^2+x+1}=\lim_{x\to\infty}\frac{6x+2}{10x+1}=\lim_{x\to\infty}\frac{6}{10}=\frac{3}{5}.$$

27. $\lim\limits_{x\to\infty}3x^4=\infty$ and $\lim\limits_{x\to\infty}5x^3+6=\infty$. Therefore, $\lim\limits_{x\to\infty}\dfrac{3x^4}{5x^3+6}$ is of the indeterminate form $\dfrac{\infty}{\infty}$.

Applying L'Hopital's Rule: $\lim\limits_{x\to\infty}\dfrac{3x^4}{5x^3+6}=\lim\limits_{x\to\infty}\dfrac{12x^3}{15x^2}=\lim\limits_{x\to\infty}\dfrac{4}{5}x=\infty$.

29. *One possible answer:* Consider the case $\dfrac{0}{0}$. We know that 1) $\lim\limits_{h\to0}\dfrac{h}{c}=0$ for any non-zero real number and 2) $\lim\limits_{h\to0}\dfrac{c}{h}$ is increasing (or decreasing) without bound for any non-zero real number. If we apply these two statements repeatedly with c approaching 0, we end with an apparent contradiction. Similar arguments can be applied for $\dfrac{\infty}{\infty}$ and $0\cdot\infty$.

Chapter 3 Concept Review

1. **a.** $x\approx 0.8$

   b. positive slope: $0.8<x<2$
   negative slope: $-3<x<0,\ 0<x<0.8$

   c. $x=0$

   d. $f'(-2)\approx -4, f'(1)\approx 1.1$, See Answer Key page A-31 in Text for figure.

   e. See Answer Key page A-31 in Text for figure.

2. **a.** Note that D is a logistic function with $L=8.101$, $A=214.8$, and $B=0.797$ and derivative of the form $D'(t)=\dfrac{LABe^{-Bt}}{(1+Ae^{-Bt})^2}=\dfrac{8.101(214.8)(0.797)e^{-0.797t}}{(1+214.8e^{-0.797t})^2}$ pounds per person per year t years after 1980

   b. $D'(10)\approx 0.4$ pound per person per year
   In 1990 the average annual per capita consumption of turkey in the United States was increasing by 0.4 pound per person per year.

   c. $D'(21)\approx 0.00007$ pound per person per year. There was essentially no growth in the per capita consumption of turkey in 2001.

3. a. $\dfrac{6890-4865}{2000-1996}=506.3$ billion per year

b.

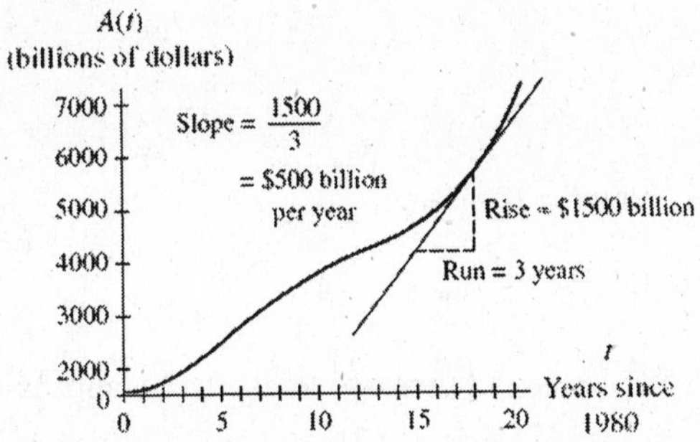

A(t)
(billions of dollars)
Slope = $\frac{1500}{3}$
= \$500 billion per year
Rise = \$1500 billion
Run = 3 years
t
Years since 1980

c. $A'(18)\approx \$496.4$ billion per year

4. a. Rewrite $P(t)$ as $P(t)=100N(t)[A(t)]^{-1}$.
Use the Product Rule to find the derivative.

$$P'(t)=100\frac{d}{dt}\left(N(t)[A(t)]^{-1}\right)$$

$$=100\left(\left[\frac{d}{dt}N(t)\right][A(t)]^{-1}+N(t)\frac{d}{dt}[A(t)]^{-1}\right)$$

$$=100\left(N'(t)[A(t)]^{-1}+N(t)(-1)[A(t)]^{-2}A'(t)\right)$$

$$=\frac{100N'(t)}{A(t)}-\frac{100N(t)A'(t)}{[A(t)]^{2}}$$ percentage points per year t years after 1980

b. Input units: years
Output units: percentage points per year

Chapter 4
Analyzing Change: Applications of Derivatives

Section 4.1 Approximating Change

1. $32\% - (4 \text{ percentage points per hour})\left(\frac{1}{3}\text{hour}\right) = 30\frac{2}{3}\%$

3. $f(3.5) \approx f(3) + f'(3)(0.5) = 17 + 4.6(0.5) = 19.3$

5. a. Increasing production from 500 to 501 units will increase total cost by approximately \$17.

   b. If sales increase from 150 to 151 units, then profit will increase by approximately \$4.75.

7. A marginal profit of –\$4 per shirt means that at this point the fraternity's profit is decreasing by \$4 for each additional shirt sold. The fraternity should consider selling fewer shirts or increasing the sales price.

9. Premium (dollars)
12,000
10,000
8000
6000
4000
2000
0
30
70
Rise ≈ \$8600
Run ≈ 10 years
Age (years)

Using the two points (70,8000 and 54,0)

$$\text{Slope of tangent line} \approx \frac{\$8000}{16 \text{ years of age}} = \$500 \text{ per year of age}$$

Annual premium for 70-year-old $\approx \$8000$

Annual premium for 72-year-old $\approx \$8000 + \left(\frac{\$500}{\text{year}}\right)(2 \text{ years}) = \$9000$

(Estimates will vary.)

11. a.

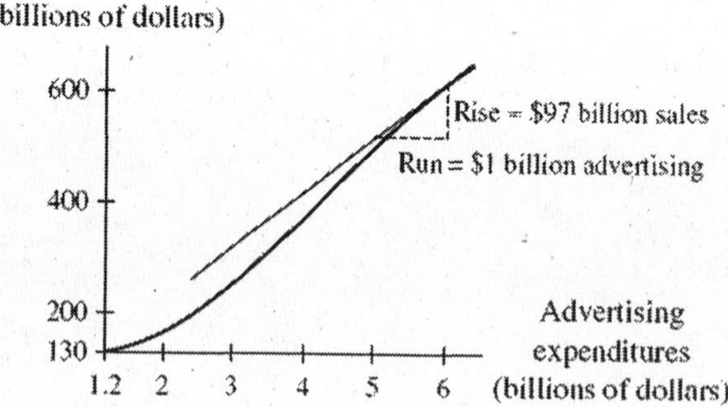

$$\text{Slope of tangent line is approximately } \frac{97 \text{ billion dollars}}{1 \text{ billion dollars}} = 97 \text{ billion dollars per billion dollars}$$

(revenue dollars per sales dollars).
Revenue is approximately \$614 billion when \$6 billion is spent on advertising.

Revenue is approximately $614+0.5(97)=662.5$ billion dollars when \$6.5 billion is spent on advertising. (Discussion will vary.)

b. $R(6.5)\approx \$658$ billion

c. The solution from the model is more accurate than that which is derived from an interpretation of the graph, because it is difficult to accurately draw a tangent line on so small a graph.

13. a. $P(x)=268.79(1.013087^x)$ thousand people in year x

$P'(x)=268.79(\ln 1.013087)(1.013087^x)$ thousand people per year in year x
In 2000 the population of South Carolina was increasing by 53.6 thousand people per year.

b. Between 2000 and 2003, the population increased by approximately 160.8 thousand people.

c. By finding the slope of the tangent line at 2000 and multiplying by 3, we determine the change in the tangent line from 2000 through 2003 and use that change to estimate the change in the population function.

15. a. The population was growing at a rate of 2.52 million people per year in 1998.

b. Between 1998 and 1999, the population of Mexico increased by approximately 2.52 million people.

17. a. In 1998 the amount was increasing by 1.15 million pieces per year.

b. We would expect an increase of approximately 1.15 million pieces between 1998 and 1999.

c. $p(24)-p(23)\approx 1.3$ million pieces

d. $101.9-100.4=1.5$ million pieces

e. As long as the data in part *d* were correctly reported, the answer to part *d* is the most accurate one.

19. a. $A=300(1+\frac{.065}{12})^{(12t)}$

b. $A=300(1.06697)^t$
c. $A(2)=\$341.53$
d. $A'(2)=300(\ln 1.06697)(1.06697)^2=22.14$ dollars/year.
e. $A(2.25)\cong A(2)+.25A'(2)=\$347.07$

21. Sales

a. $R(x)=\left(-7.032\cdot 10^{-4}\right)x^2+1.666x+47.130$ dollars when x hot dogs are sold, $100<x<1500$.

b. Cost: $C(x)=0.5x$ dollars when x hot dogs are sold, $100<x<1500$.

Profit: $P(x)=R(x)-C(x)=(-7.032\cdot 10^{-4})x^2+1.166x+47.130$ dollars when x hot dogs are sold, $100<x<1500$.

c. Marginal Revenue = $R'(x) = -.0014x + 1.666$ dollars/hot dog.

x (hot dogs)	$R'(x)$ (dollars per hot dog)	$c'(x)$ (dollars per hot dog)	$p'(x)$ (dollars per hot dog)
200	1.38	0.50	0.88
800	0.54	0.50	0.04
1100	0.12	0.50	–0.38
1400	–0.30	0.50	–0.80

If the number of hot dogs sold increases from 200 to 201, the revenue increases by approximately \$1.38 and the profit increases by approximately \$0.88. If the number increases from 800 to 801, the revenue increases by 0.54, but the profit sees almost no increase (4 cents). If the number increases from 1100 to 1100, the increase in revenue is only approximately 12 cents. Because this marginal revenue is less than the marginal cost at a sales level of 1100, the result of the sales increase from 1100 to 1101 is a decrease of \$0.38 in profit. If the number of hot dogs increases from 1400 to 1401, revenue declines by approximately 30 cents and profit declines by approximately 80 cents.

d.

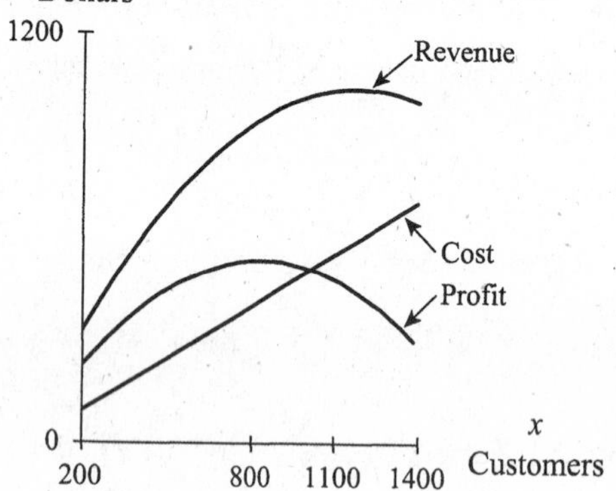

The marginal values in part *c* are the slopes of the graphs shown here. For example, at $x = 800$, the slope of the revenue graph is \$0.54 per hot dog, the slope of the cost graph is \$0.50 per hot dog, and the slope of the profit graph is \$0.04 per hot dog. We see from the graph that maximum profit is realized when approximately 800 hot dogs are sold.

Revenue is greatest near $x = 1100$, so the marginal revenue there is small. However, once costs are factored in, the profit is actually declining at this sales level. This is illustrated by the graph.

23. CPI

United States:

a. $A(t) = 0.109t^3 - 1.555t^2 + 10.927t + 100.320$ t years after 1980

b. $A'(t) = 0.327t^2 - 3.111t + 10.927$ index points per year t years after 1980

$A'(7) \approx 5.2$ index points per year

c. 1988 CPI estimate:

(CPI in 1987) + $A'(7)$ (1 year) $\approx$ 137.9 + (5.2 index points per year)(1 year)

$= 143.1$

Note: The estimate can also be calculated using the value of $A(7)$ instead of the actual CPI in 1987. Because the model closely agrees with the actual value in 1987, the value of this estimate is not significantly affected by this choice.

Canada:

a. $C(t) = 0.150t^3 - 2.171t^2 + 15.814t + 99.650$ t years after 1980

b. $C'(t) = 0.450t^2 - 4.343t + 15.814$ index points per year t years after 1980
$C'(7) \approx 7.5$ index points per year

c. 1988 CPI estimate:
(CPI in 1987) + $C'(7)$ (1 year) $\approx$ 155.4 + (7.5 index points per year)(1 year)
= 162.9

Note: The estimate can also be calculated using the value of $C(7)$ instead of the actual CPI in 1987. Because the model closely agrees with the actual CPI in 1987, the value of this estimate is not significantly affected by this choice.

Peru:

a. $P(t) = 85.112(2.013252^t)$ t years after 1980

b. $P'(t) = 85.112(\ln 2.013252)(2.013252^t)$

$\approx 59.558(2.013252^t)$ index points per year t years after 1980
$P'(7) \approx 7984$ index points per year

c. 1988 CPI estimate:
(CPI in 1987) + $P'(7)$ (1 year) $\approx$ 11,150 + (7984 index points per year)(1 year)
= 19,134

Note: The estimate can also be calculated using the value of $P(7)$ instead of the actual CPI in 1987. If this is done, the estimate will be approximately 19,394.

Brazil:

a. $B(t) = 73{,}430(2.615939^t)$ t years after 1980

b. $B'(t) = 73.430(\ln 2.615939)(2.615939^t)$

$\approx 70.612(2.615939^t)$ index points per year t years after 1980
$B'(7) \approx 59{,}193$ index points per year

c. 1988 CPI estimate:
(CPI in 1987) + $B'(7)$ (1 year) $\approx$ 77,258 + (59,193 index points per year)(1 year)
= 136,451

Note: The estimate can also be calculated using the value of $B(7)$ instead of the actual CPI in 1987. If this is done, the estimate will be approximately 120,748.

25. Advertising

Note: This Activity can be solved using either a cubic model or a logistic model. The following solution uses a cubic model.

a. $R(A) = -0.158A^3 + 5.235A^2 - 23.056A + 154.884$ thousand dollars of revenue when A thousand dollars is spent on advertising. $5 < A < 19$.

b. $R'(A) = -0.473A^3 + 10.471A^2 - 23.056$ thousand dollars of revenue per thousand dollars of advertising when A thousand dollars is spent on advertising
$R'(10) \approx 34.3$ thousand dollars of revenue per thousand dollars of advertising
When $10,000 is spent on advertising, revenue is increasing by $34.3 thousand per thousand advertising dollars. If advertising is increased from $10,000 to $11,000, the car dealership can expect an approximate increase in revenue of $34,300.

c. $R'(18) \approx 12.0$ thousand dollars of revenue per thousand of dollars of advertising
When \$18,000 is spent on advertising, revenue is increasing by \$12.0 thousand per thousand advertising dollars. If advertising is increased from \$18,000 to \$19,000, the car dealership can expect an approximate increase in revenue of \$12,000.

27. *One possible answer:* Close to the point of tangency, a tangent line and a curve are close to one another. The farther away from the point of tangency we move, the more the tangent line deviates from the curve. Thus the tangent line near the point of tangency will usually produce a good estimate, but the tangent line farther away from the point of tangency will produce a poor estimate.

29. *One possible answer:* By definition $f'(x) = \lim_{h \to 0} \frac{f(x+h) - f(x)}{h}$. Assuming h is relatively close to zero, $f'(x) \approx \frac{f(x+h) - f(x)}{h}$. Multiplying both sides of this approximation by h yields $h \cdot f'(x) \approx f(x+h) - f(x)$.

Section 4.2 Relative and Absolute Extreme Points

1. Quadratic, cubic, and many product, quotient, and composite functions could have relative maxima or minima.

3.

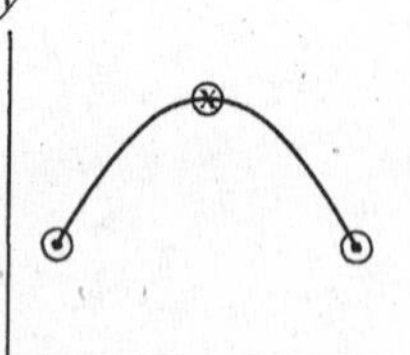

The derivative is zero at the absolute maximum point.

5.

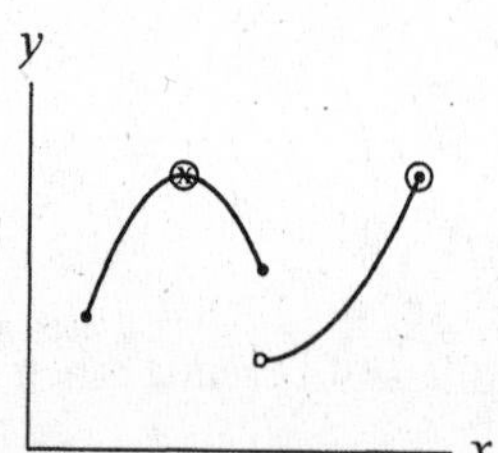

The derivative is zero at the absolute maximum point marked with an X. The derivative where the graph is broken, at the relative minimum, is undefined.

7.

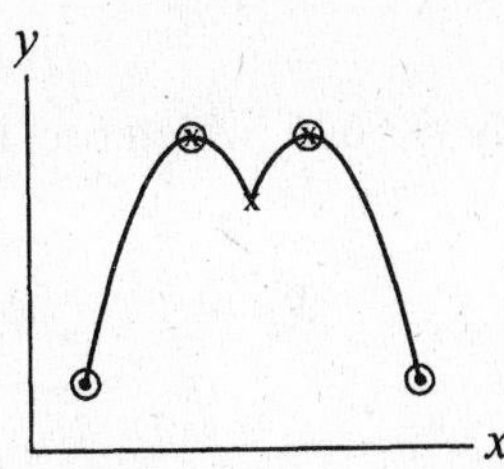

The derivative is zero at both absolute maximum points. The derivative does not exist at the relative minimum point.

9. *One possible answer:* One such graph is $y = x^3$, which does not have a relative minimum or maximum at $x = 0$ even though the derivative is zero at this point.

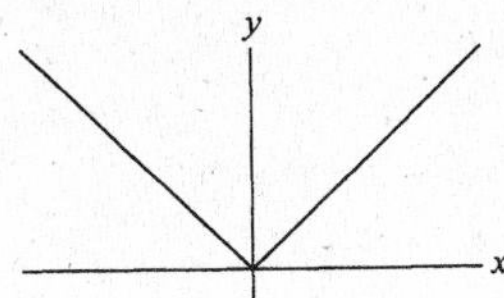

11. **a.** All statements are true.

b. The derivative does not exist at
$x = 2$ because f is not continuous there, so the third statement is false.

c. The slope of the graph is negative, $f'(x) < 0$, to the left of $x = 2$ because the graph is decreasing, so the second statement is false.

d. The derivative does not exist at
$x = 2$ because f is not smooth there, so the third statement is not true.

13. *One possible graph:*

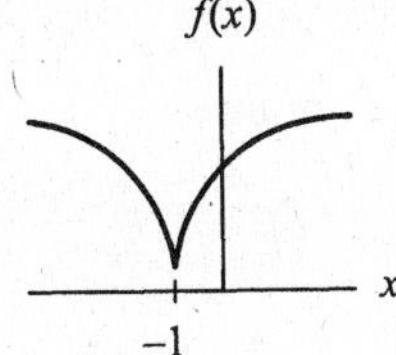

15. *One possible graph:*

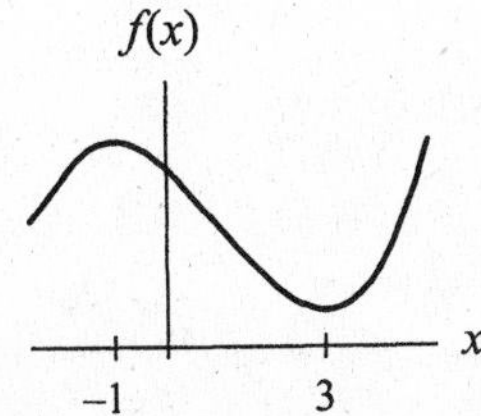

17. a. The derivative formula is $f'(x) = 2x + 2.5$

b. Using technology, the relative minimum value is approximately -7.5625, which occurs at $x \approx$ -1.25.

c. 

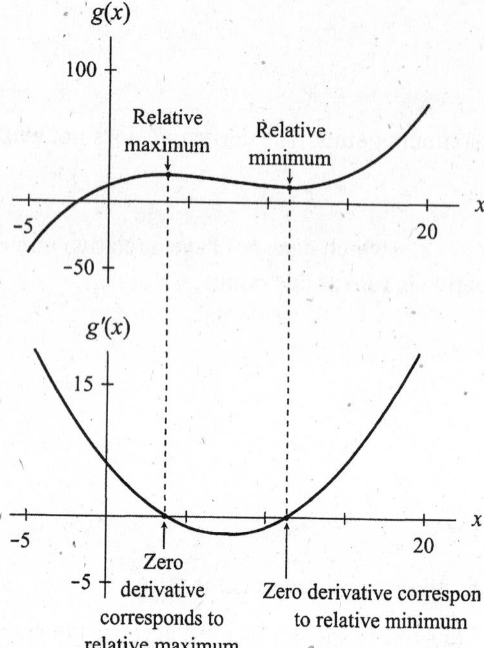

19. a. $h'(x) = 3x^2 - 16x - 6$

b. The relative maximum is 1.077, which occurs at $x \approx$ -0.352;
The relative minimum is -108.929, which occurs at $x \approx$ 5.685

21. a. $f'(t) = 12(\ln 1.5)(1.5^t) + 12(\ln 0.5)(0.5^t)$

b. This function is always increasing, so does not have a relative maximum nor relative minimum.

23. a. $g'(x) = .12x^2 - 1.76x + 4.81$

b. The relative maximum is 19.888, which occurs at $x \approx$ 3.633;
The relative minimum is 11.779, which occurs at $x \approx$ 11.034

c. On the closed interval [0 , 14.5],
The absolute minimum is found at the point (11.034, 11.779);
The absolute maximum is found at the point (3.633, 19.888)

25. Grasshoppers

a. At 9.449 °C, the greatest percentage of eggs, 95.598%, eggs hatch.

b. 9.449 °C corresponds to 49 °F.

27. River Rate

a. The flow rate for $h = 0$ was 123.02 cfs; for $h = 11$ it was 331.305 cfs.

b. The absolute minimum is found at the point (.388, 121.311), or when $h \approx 0.4$ hours.
The absolute maximum is found at the point (8.900, 387.975) or $h \approx 8.9$ hours.

29. Swim Time

a. $S(x) = 0.181x^2 - 8.463x + 147.376$ seconds at age x years.

b. The model gives a minimum time of 48.5 seconds occurring at 23.4 years.

c. The minimum time in the table is 49 seconds, which occurs at 24 years of age.

31. Sales

a. A quadratic or exponential model can be used to model the data, but the exponential model may be a better choice because it does not predict that demand will increase for prices above \$40. An exponential model for the data is $R(p) = 316.765(0.949^p)$ dozen roses when the price per dozen is p dollars.

b. Multiply $R(p)$ by the price, p. The consumer expenditure is $E(p) = 316.765p(0.949^p)$ dollars spent on roses each week when the price per dozen is p dollars.

c. Using technology, we find that $E(p)$ is maximized at $p \approx 19.16$ dollars. A price of \$19.16 per dozen maximizes consumer expenditure.

d. Profit is given by $F(p) = E(p) - 6R(p) = 316.765(p-6)(0.949^p)$.
Using technology, $F(p)$ is maximized at $p \approx 25.16$ dollars. A price of \$25.16 per dozen maximizes profit.

e. Marginal values are with respect to the number of units sold or produced. In this activity, the input is price, so derivatives are with respect to price and are not marginals.

33. Refuse

a. $G(t) = 0.008t^3 - 0.347t^2 + 6.108t + 79.690$ million tons of garbage taken to a landfill t years after 1975

b. $G'(t) = 0.025t^2 - 0.693t + 6.108$ million tons of garbage per year t years after 1975

c. In 2005 the amount of garbage was increasing by $G'(30) \approx 8.1$ million tons per year.

d.

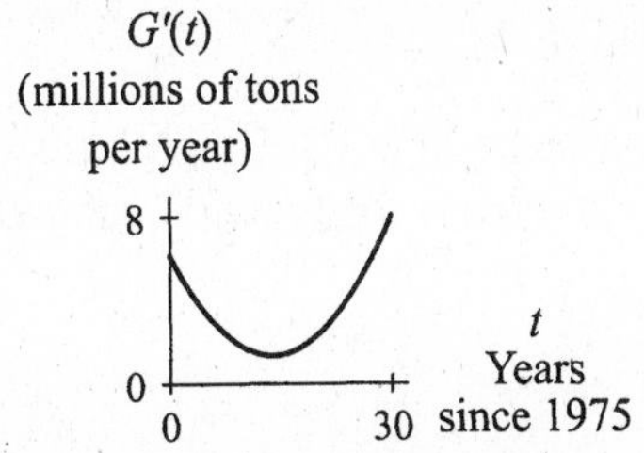

$G'(t)$
(millions of tons per year)
8
0
0
30
t
Years since 1975

Because the derivative graph exists for all input values and never crosses the horizontal axis, $G(t)$ has no relative maxima.

35. *One possible answer:* The graph shown below indicates there is an absolute maximum to the right of -3 and an absolute minimum to the left of 1. A view of the graph showing more of the horizontal axis indicates that $y = 2$ is a horizontal asymptote for the graph.

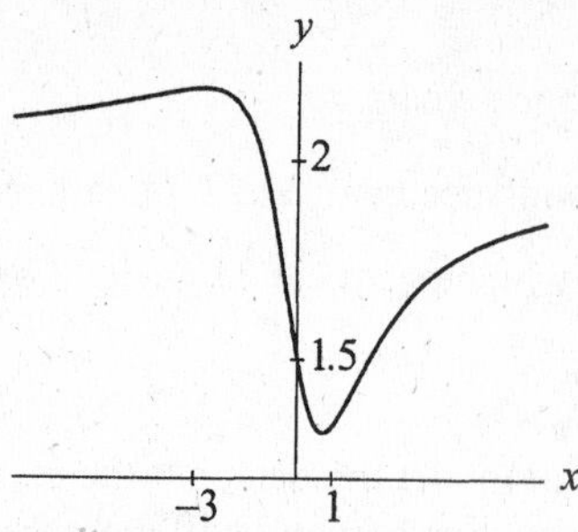

Use technology to find the absolute extrema, or solve the equation

$$y' = \frac{-2x(2x^2 - x + 3)}{(x^2+2)^2} + \frac{4x-1}{x^2+2} = 0$$

In either case you should find the absolute minimum point of approximately (0.732, 1.317) and the absolute maximum point of approximately (–2.732, 2.183). Thus the absolute maximum is approximately 2.18, and the absolute minimum is approximately 1.32.

Section 4.3 Inflection Points

1. **Production**

   a. One visual estimate of the inflection points is (1982, 25) and (2018, 25). *Note:* There are also "smaller" inflection points at approximately (1921, 2.5), (1927, 2), (1930, 2), and (1935, 3).

   b. The input values of the inflection points are the years in which the rate of crude oil production is estimated to be increasing and decreasing most rapidly. We estimate that the rate of production was increasing most rapidly in 1982, when production was approximately 25 billion barrels per year, and that it will be decreasing most rapidly in

3. For polynomial functions, as these appear to be, you can identify the function and its derivative by noticing the number of inflection points. Because a derivative has a power one less than the original function, it will also have one less inflection point. Thus graph *b* with two inflection points is the function. Graph *a* with one inflection point is the derivative, and graph *c* with no inflection points is the second derivative.

5. Graph *c* appears to have a minimum at –1 and an inflection point at –2. Graph *b* crosses the horizontal axis at –1 and graph *a* crosses it at –2. Thus graph c is the function, graph *b* is the derivative, and graph *a* is the second derivative.

7. $f'(x) = -3$; $f''(x) = 0$

9. $c'(u) = 6u - 7$; $c''(u) = 6$

11. $p'(u) = -6.3u^2 + 7u$; $p''(u) = -12.6u + 7$

13. $g'(t) = 37(\ln 1.05)(1.05^t)$; $g''(t) = 37(\ln 1.05)^2(1.05^t)$

15. $f'(x) = 3.2x^{-1}$; $f''(x) = -3.2x^{-2}$

17. $L'(t) = -131.04e^{3.9t}(1+2.1e^{3.9t})^{-2}$; L"(t) = $L''(t) = \dfrac{-2146.4352e^{7.8t}}{\left(1+2.1e^{3.9t}\right)^3} + \dfrac{511.056e^{3.9t}}{\left(1+2.1e^{3.9t}\right)^2}$

19. $f'(x) = 3x^2 - 12x + 2$; $f''(x) = 6x - 12 = 0$ at $x = 2$

21. Using technology, the second derivative = 0 at the point x = 3.356

23. There is no value for t for which the second derivative is zero, therefore there is no inflection point.

25. a. $g(x) = 0.04x^3 - 0.88x^2 + 4.81x + 12.11$

$g'(x) = 0.12x^2 - 1.76x + 4.81$

$g''(x) = 0.24x - 1.76$

b. The inflection point on the graph of g is approximately (7.333, 15.834). This is a point of most rapid decline.

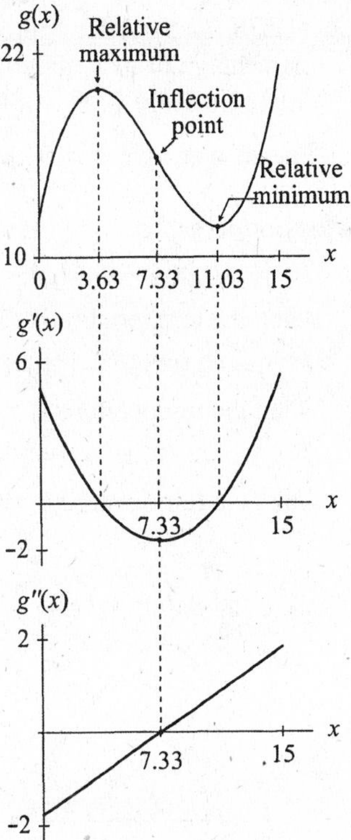

27. Study Time

a. $P(t) = \dfrac{45}{1 + 5.94e^{-0.969125t}} = 45\left(1 + 5.94e^{-0.969125t}\right)^{-1}$ percent after studying for t hours

$$P'(t) = 45(-1)\left(1 + 5.94e^{-0.969125t}\right)^{-2}\left(5.94e^{-0.969125t}\right)(-0.969125)$$

$$= 259.0471125e^{-0.969125t}\left(1 + 5.94e^{-0.969125t}\right)^{-2}$$ percentage points per hour after studying for t hours

$$P''(t) = 259.0471125\left(\frac{d}{dx}\left(e^{-0.969125t}\right)\right)\left(1 + 5.94e^{-0.969125t}\right)^{-2}$$

$$+ 259.0471125e^{-0.969125t}\left(\frac{d}{dx}\left(1 + 5.94e^{-0.969125t}\right)^{-2}\right)$$

$$= 259.0471125\left(e^{-0.969125t}\right)(-0.969125)\left(1 + 5.94e^{-0.969125t}\right)^{-2}$$

$$+ 259.0471125e^{-0.969125t}(-2)\left(1 + 5.94e^{-0.969125t}\right)^{-3}\left(5.94e^{-0.969125t}\right)(-0.969125)$$

$$\approx -251.049033e^{-0.969125t}\left(1+5.94e^{-0.969125t}\right)^{-2}$$

$$+2982.462511e^{-1.93825t}(1+5.94e^{-0.969125t})^{-3}$$ percentage points per hour per hour after studying for t hours

Solving $P''(t)=0$ for t gives $t \approx 1.838$. The inflection point on P is approximately (1.838, 22.5). After approximately 1.8 hours of study (1 hour and 50 minutes), the percentage of new material being retained is increasing most rapidly. At that time, approximately 22.5% of the material has been retained.

b. The answer agrees with that given in the discussion at the end of the section.

29. Grasshoppers

a. $P(t) = -0.00645t^4 + 0.488t^3 - 12.991t^2 + 136.560t - 395.154$ percent when the temperature is t°C

$P'(t) = -0.0258t^3 + 1.464t^2 - 25.982t + 136.560$ percentage points per °C when the temperature is t°C

$P''(t) = -0.0774t^2 + 2.928t - 25.982$ percentage points per °C per °C when the temperature is t°C

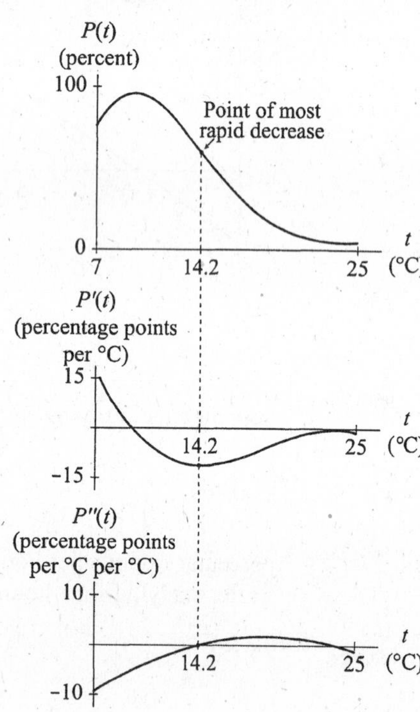

b. Because the graph of P'' crosses the t-axis twice, there are two inflection points. These are approximately (14.2, 59.4) and (23.6, 5.8). The point of most rapid decrease on the graph of P is (14.2, 59.4). (The other inflection point is a point of least rapid decrease.) The most rapid decrease occurs at 14.2°C, when 59.4% of eggs hatch. At this temperature, $P'(14.2) \approx -11.1$, so the percentage of eggs hatching is declining by 11.1 percentage points per °C. A small increase in temperature will result in a relatively large increase in the percentage of eggs not hatching.

31. Price

a. The relative maximum point on the derivative graph between the values x = 4 and x = 10 occurs where x ≅ 5.785. This is the inflection point on the original function. The relative minimum point on the derivative graph between the values x = 4 and x = 10 occurs where x ≅ 8.115. This is the other inflection point on the original function.

b. The relative max and min on the first derivative graph correspond to the x-intercepts on the second derivative graph.
c. According to the model, between 1990 and 2001, the gas prices were declining most rapidly in 1990, and they were increasing most rapidly in 2001. This is easiest to see if one examines the first derivative graph.
d. According to the model, between 1994 and 2000, the gas price was decreasing most rapidly in 1999 (approximately where $x = 8.115$) and it was increasing most rapidly in 1996 (approximately where $x = 5.785$).

33. Donors

$D(t) = -10.247t^3 + 208.114t^2 - 168.805t + 9775.035$ donors t years after 1975

$D'(t) = -30.741t^2 + 416.288t - 168.805$ donors per year t years after 1975

a. Using technology, we find that (0.418, 9740.089) is the approximate relative minimum point, and (13.121, 20,242.033) is the approximate relative maximum point on the cubic model.

b. The inflection point occurs where $D'(t)$ has its maximum, at $t \approx 6.770$. The inflection point is approximately (6.8, 14,991.1).

c. i. Because 6.8 is between $t = 6$ (the end of 1981) and $t = 7$ (the end of 1982), the inflection point occurs during 1982, shortly after the team won the National Championship. This is when the number of donors was increasing most rapidly.

ii. The relative maximum occurred around the same time that a new coach was hired. After this time, the number of donors declined.

35. Labor

a. $N(h) = \dfrac{62}{1 + 11.49e^{-0.654h}}$ components after h hours. Using the formula $\dfrac{LABe^{-Bx}}{(1 + Ae^{-Bx})^2}$ for the derivative, we have

$$N'(h) = \frac{62(-0.654)(11.49)\left(e^{-0.654h}\right)}{\left(1 + 11.49e^{-0.654h}\right)^2} = 465.89652e^{-0.654h}\left(1 + 11.49e^{-0.654h}\right)^{-2}$$

components per hour after h hours.

The greatest rate occurs when $N'(h)$ is maximized, at $h \approx 3.733$ hours, or approximately 3 hours and 44 minutes after she began working.

b. Her employer may wish to give her a break after 4 hours to prevent a decline in her productivity.

37. Labor

a. $H(w)=\dfrac{10{,}111.102}{1+1153.222e^{-0.727966w}}$ total labor-hours after w weeks

b. $H'(w)=10{,}111.102(-1)\left(1+1153.222e^{-0.727966w}\right)^{-2}\left(1153.22e^{-0.727966w}\right)(-0.727966)$

$\approx 8{,}488{,}330.433e^{-0.727966w}\left(1+1153.222e^{-0.727966w}\right)^{-2}$ labor-hours per week after w weeks

c.

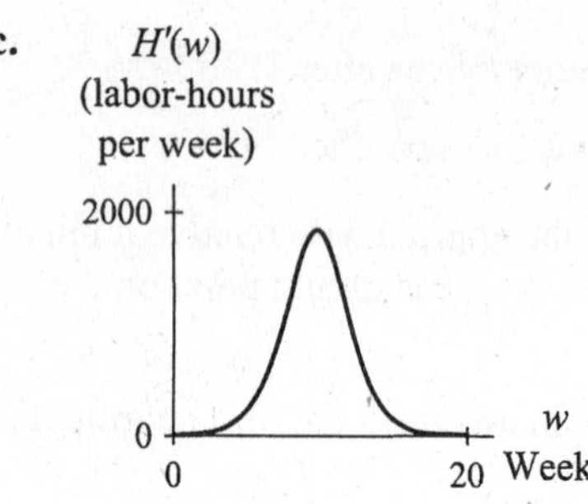

The derivative gives the manager information approximately the number of labor-hours spent each week

d. The maximum point on the graph of H' is approximately (9.685, 1840.134). Keeping in mind that the model must be discretely interpreted, we conclude that in the tenth week, the most labor-hours are needed. That number is $H'(10)\approx 1816$ labor-hours.

e. $H''(w)=8{,}488{,}330.433\left(\dfrac{d}{dx}\left(e^{-0.727966w}\right)\right)\left(1+1153.222e^{-0.727966w}\right)^{-2}$

$+8{,}488{,}330.433\left(e^{-0.727966w}\right)\left(\dfrac{d}{dx}\left(1+1153.222e^{-0.727966w}\right)^{-2}\right)$

$=8{,}488{,}330.433\left(e^{-0.727966w}\right)(-0.727966)\left(1+1153.222e^{-0.727966w}\right)^{-2}$

$+8{,}488{,}330.433\left(e^{-0.727966w}\right)(-2)\left(1+1153.222e^{-0.727966w}\right)^{-3}$

$\left(1153.222e^{-0.727966w}\right)(-0.727966)$

$\approx -6{,}179{,}214.512\left(e^{-0.727966w}\right)\left(1+1153.222e^{-0.727966w}\right)^{-2}$

$+\left(1.4252\cdot 10^{10}\right)\left(e^{-1.455932w}\right)\left(1+1153.222e^{-0.72766w}\right)^{-3}$

Use technology to find the maximum of $H''(w)$, which occurs at $w\approx 7.876$. The point of most rapid increase on the graph of H' is (7.876, 1226.756). This occurs approximately 8 weeks into the job, and the number of labor-hours per week is increasing by approximately $H''(8)\approx 513$ labor-hours per week per week.

f. Use technology to find the minimum of the graph of H'', which occurs at $w\approx 11.494$. The point of most rapid decrease on the graph of H' is (11.494, 1226.756). This occurs approximately 12 weeks into the job when the number of labor-hours per week is changing by approximately $H''(12)=-486$ labor-hours per week per week.

g. By solving the equation $H'''(w) = 0$, we can find the input values that correspond to a maximum or minimum point on the graph of H'', which corresponds to inflection points on the graph of H', the weekly labor-hour curve.

h. Since the minimum of $H''(w)$ occurs approximately 4 weeks after the maximum of $H''(w)$, the second job should begin approximately 4 weeks into the first job.

39. Refuse

a. Between 1980 and 1985, the average rate of change was smallest at
$\frac{122 - 117}{1985 - 1980} = 1$ million tons per year.

b. $g(t) = 0.008t^3 - 0.347t^2 + 6.108t + 79.690$ million tons t years after 1970

c. $g'(t) = 0.025t^2 - 0.693t + 6.108$ million tons per year t years after 1970
$g''(t) = 0.051t - 0.693$ million tons per year per year t years after 1970

d.
$$g''(t) = 0$$
$$0.0507t - 0.693 = 0$$
$$0.0507t = 0.693$$
$$t \approx 13.684$$

Solving $g''(t) = 0$ gives $t \approx 13.684$, which corresponds to mid-1984. The corresponding amount of garbage is $g(13.684) \approx 120$ million tons and the corresponding rate of increase is $g'(13.684) \approx 1.4$ million tons per year.

e.

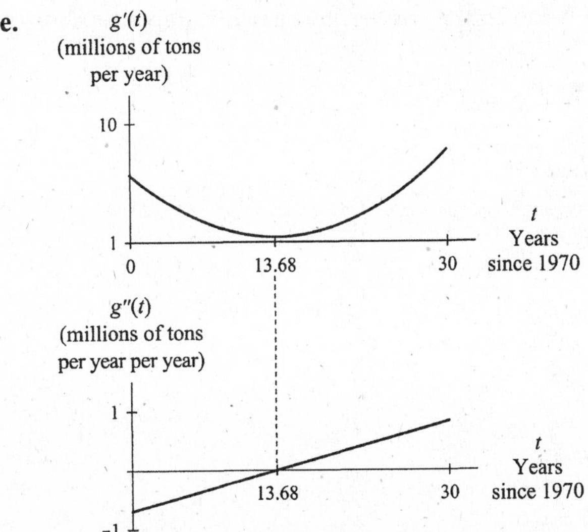

Because the graph of g'' crosses the t-axis at 13.68, we know that input corresponds to an inflection point of the graph of g. Because the graph of g' has a minimum at that same value, we know that it corresponds to a point of slowest increase on the graph of g.

f. The year with the smallest rate of change is 1984, with $g(14) \approx 120.4$ million tons of garbage, increasing at a rate of $g'(14) \approx 1.37$ million tons per year.

41. Reaction

a. The first differences are greatest between 6 and 10 minutes, indicating the most rapid increase in activity.

b. $A(m) = \dfrac{1.930}{1+31.720e^{-0.439118m}}$ U / 100 μL *m* minutes after the mixture reaches 95°C

The inflection point, whose input is found using technology to locate the maximum point on a graph of A', is approximately (7.872, 0.965). After approximately 7.9 minutes, the activity was approximately 0.97 U/100 μL and was increasing most rapidly at a rate of approximately 0.212 U/100 μL/min.

43. The graph of f is always concave up. A parabola that opens upward fits this description.

45. a. The graph is concave up between $x = 0$ and $x = 2$, has an inflection point at $x = 2$ and is concave down between $x = 2$ and $x = 4$.

b.

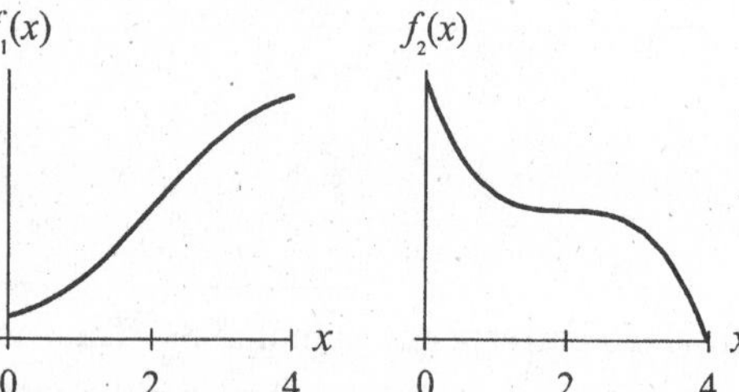

47. *One possible answer:* Cubic and logistic models have inflection points, as do some product, quotient, and composite functions. **Note: In the answer in the text exponential is listed twice and cubic is omitted.**

Section 4.4 Interconnected Change: Related Rates

1. $\dfrac{df}{dt} = 3\dfrac{dx}{dt}$

3. $\dfrac{dk}{dy} = 12x\dfrac{dx}{dy}$

5. $\dfrac{dg}{dt} = 3e^{3x}\dfrac{dx}{dt}$

7. $\dfrac{df}{dt} = 62(\ln 1.02)(1.02^x)\dfrac{dx}{dt}$

9. $\frac{dh}{dy} = 6\frac{da}{dy} + 6\ln a\frac{da}{dy}$

$= 6(1+\ln a)\frac{da}{dy}$

11. $\frac{ds}{dt} = \pi r\frac{1}{2}\left(r^2+h^2\right)^{-1/2}(2h)\frac{dh}{dt}$

$= \frac{\pi r h}{\sqrt{r^2+h^2}}\frac{dh}{dt}$

13. Use the Product Rule with πr as the first term and $\sqrt{r^2+h^2}$ as the second term.

$0 = \pi r\frac{1}{2}\left(r^2+h^2\right)^{-1/2}\left(2r\frac{dr}{dt}+2h\frac{dh}{dt}\right)+\left(\pi\frac{dr}{dt}\right)\sqrt{r^2+h^2}$

$0 = \frac{\pi r}{\sqrt{r^2+h^2}}\left(r\frac{dr}{dt}+h\frac{dh}{dt}\right)+\pi\sqrt{r^2+h^2}\frac{dr}{dt}$

15. a. $w = 31.54+12.97\ln 5 \approx 52.4$ gallons per day

b. $\frac{dw}{dt} = \frac{12.97}{g}\frac{dg}{dt} = \frac{12.97}{5}\left(\frac{2}{12}\text{ inches per year}\right) \approx 0.43$ gallon per day per year

The amount of water transpired is increasing by approximately 0.43 gallon per day per year. In other words, in one year, the tree will be transpiring approximately 0.4 gallon more each day than it currently is transpiring.

17. a. $B = \frac{0.45(100)}{0.00064516h^2} = \frac{45}{0.00064516h^2}$ points

b. $\frac{dB}{dt} = \frac{45}{0.00064516}\left(-2h^{-3}\right)\frac{dh}{dt} = \frac{45}{0.00064516h^3}\frac{dh}{dt}$

c. Evaluate the equation in part *b* at $h = 63$ inches and $\frac{dh}{dt} = 0.5$ inch per year to obtain

$\frac{dB}{dt} \approx -0.2789$ point per year.

19. a. We know $h = 32$ feet, $d = \frac{10}{12}$ foot, $\frac{dh}{dt} = 0.5$ foot per year, and we wish to find $\frac{dV}{dt}$. We treat d as a constant and find the derivative with respect to time t to obtain the related rates equation $\frac{dV}{dt} = 0.002198d^{1.739925}1.133187h^{0.133187}\frac{dh}{dt}$.

Substituting the values given above results in $\frac{dV}{dt} \approx 0.0014$ cubic foot per year.

b. We know $h = 34$ feet, $d = 1$ foot, $\frac{dd}{dt} = \frac{2}{12}$ foot per year, and we wish to find $\frac{dV}{dt}$. We treat h as a constant and find the derivative with respect to time t to obtain the related rates equation $\frac{dV}{dt} = 0.002198(1.739925d^{0.739925})h^{1.133187}\frac{dd}{dt}$.

Substituting the values given above results in $\frac{dV}{dt} \approx 0.0347$ cubic foot per year.

21. a. $L = \left(\dfrac{M}{48.10352K^{0.4}}\right)^{5/3} = \left(\dfrac{M}{48.10352}\right)^{5/3} K^{-2/3}$

b. $\dfrac{dL}{dt} = \left(\dfrac{M}{48.10352}\right)^{5/3} \left(\dfrac{-2}{3}K^{-5/3}\right)\dfrac{dK}{dt}$

c. We are given $K = 47$ and $\frac{dK}{dL} = 0.5$. Using the fact that $L = 8$ and the original equation, we can find the value of M corresponding to the current levels of labor and capital: $M \approx 781.39$. Substituting the known values into the equation in part b gives $\frac{dL}{dt} \approx -0.057$ thousand worker-hours per year.

23.

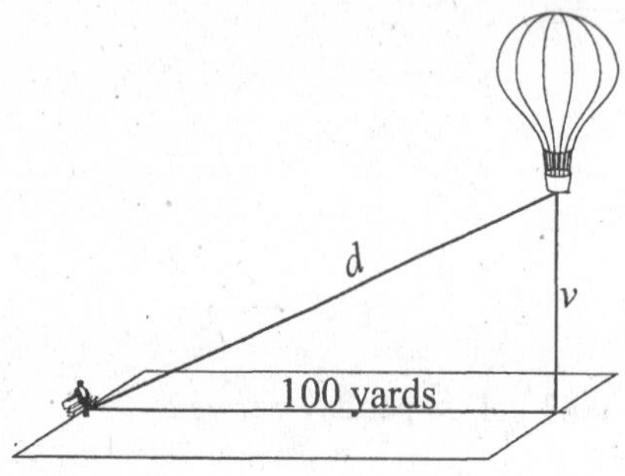

We are told that $\frac{dV}{dt} = 2$ feet per second and $v = (500 \text{ yards})(3 \text{ feet per yard}) = 1500$ feet, and we need to know $\frac{dd}{dt}$. Converting 100 yards to feet and using the Pythagorean Theorem, we know that $v^2 + 300^2 = d^2$. Taking the derivatives of both sides with respect to time gives $2v\frac{dv}{dt} + 0 = 2d\frac{dd}{dt}$.

To solve for $\frac{dd}{dt}$, we need to know the value of d when $v = 1500$ feet. Use the Pythagorean Theorem: $1500^2 + 300^2 = d^2$ to find that $d \approx 1529.71$ feet. Thus we have

$$2v\frac{dv}{dt} = 2d\frac{dd}{dt}$$

$$9(1500)(2) = 2(1529.71)\frac{dd}{dt}$$

$$\frac{dd}{dt} \approx 1.96 \text{ feet per second}$$

The balloon is approximately 1529.7 feet from the observer, and that distance is increasing by approximately 1.96 feet per second.

25.

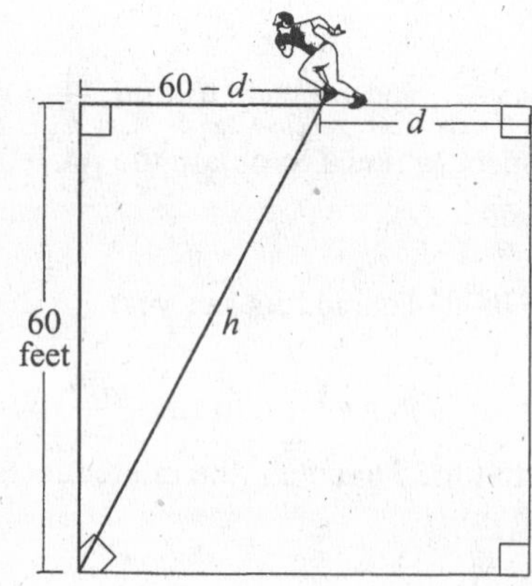

We are told that $\frac{dd}{dt} = 22$ feet per second and $d = 30$ feet. We wish to find $\frac{dh}{dt}$. We use the Pythagorean Theorem: $h^2 = 60^2 + (60 - d)^2$ to obtain the related rates equation:

$$2h\frac{dh}{dt} = 0 + 2(60 - d)(-1)\frac{dd}{dt}$$

To find the value of h, we substitute $d = 30$ into the Pythagorean Theorem: $h^2 = 60^2 + 30^2$

$$h \approx 67.08$$

Thus we have $2h\frac{dh}{dt} = -2(60-d)\frac{dd}{dt}$

$$2(67.08)\frac{dh}{dt} = -2(60-30)(22)$$

$$\frac{dh}{dt} \approx \frac{1320}{2(67.08)} \approx 9.8 \text{ feet per second}$$

The runner is approximately 67.1 feet from home plate, and that distance is decreasing by approximately 9.84 feet per second.

27. a. The volume of a sphere with radius r centimeters is given by the formula $V = \frac{4}{3}\pi r^3$ cubic centimeters. When $r = 10$, $V \approx 4188.79$ cm^3.

b. Differentiating the volume equation with respect to time t yields $\frac{dV}{dt} = \frac{4}{3}(3\pi r^2)\frac{dr}{dt}$.

We find $\frac{dr}{dt}$ as the average rate of change between the points (0, 12) and (30, 8):

$\frac{dr}{dt} = \frac{8-12}{30-0} = \frac{-4}{30}$ cm per minute

Substituting $r = 10$ and $\frac{dr}{dt} = \frac{-4}{30}$ into the related rates equation, we find that

$\frac{dV}{dt} = \frac{4}{3}(3\pi(10^2))\frac{-4}{30} \approx -167.6$ cm^3 per minute

29.

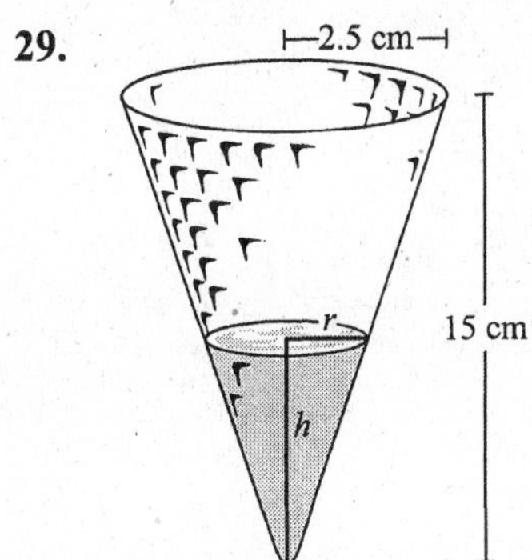

We are told $\frac{dV}{dt} = \frac{1\text{ T}}{\text{sec}} = \frac{1\text{ T}}{\text{sec}}\left(\frac{1\text{ cm}^3}{0.06\text{T}}\right) = \frac{1}{0.06}\text{cm}^3$ per second

We wish to find $\frac{dh}{dt}$. The volume of a cone with radius r and height h, both in centimeters is $V = \frac{\pi r^2 h}{3}$ cm^3. Because of similar triangles, we know that $\frac{15}{2.5} = \frac{h}{r}$ or $r = \frac{h}{6}$. Substituting this expression into the volume equation gives volume in terms of height: $V = \frac{\pi\left(\frac{h}{6}\right)^2 h}{3} = \frac{\pi h^3}{108}$

Differentiating both sides with respect to time t gives $\frac{dV}{dt} = \frac{3\pi h^2}{108}\frac{dh}{dt}$.

When $h = 6$ cm and $\frac{dV}{dt} = \frac{1}{0.06}$, $\frac{1}{0.06} = \frac{3\pi(6^2)}{108}\frac{dh}{dt}$ which gives $\frac{dh}{dt} \approx 4.34$ cm per second

31. *One possible answer:*

Begin by solving for h: $h = \frac{V}{\pi r^2} = \frac{V}{\pi} r^{-2}$

Differentiate with respect to t (V is a constant): $\frac{dh}{dt} = \frac{V}{\pi}\left(-2r^{-3}\right)\frac{dr}{dt}$

Substitute $\pi r^2 h$ for V: $\frac{dh}{dt} = \frac{\pi r^2 h}{\pi}\left(-2r^{-3}\right)\frac{dr}{dt}$

Simplify: $\frac{dh}{dt} = \frac{-2h}{r}\frac{dr}{dt}$

Rewrite: $\frac{dr}{dt} = \frac{r}{-2h}\frac{dh}{dt}$

32. *One possible answer:* The first step, wherein the independent and dependent variables are identified, is the most critical step in solving the problem correctly because if the problem is not set up properly, the correct answer will not be found.

Chapter 4 Concept Review

1. **a.** T has a relative maximum point at (0.682, 143.098) and a relative minimum point at (3.160, 120.687). These points can be determined by finding the values of x between 0 and 6 at which the graph of T' crosses the x-axis. (There is also a relative maximum to the right of $x = 6$.)

b. T has two inflection points: (1.762, 132.939) and (5.143, 149.067). These points can be determined by finding the values of x between 0 and 6 at which the graph of T'' crosses the x-axis. These are also the points at which T' has a relative maximum and relative minimum.

c.

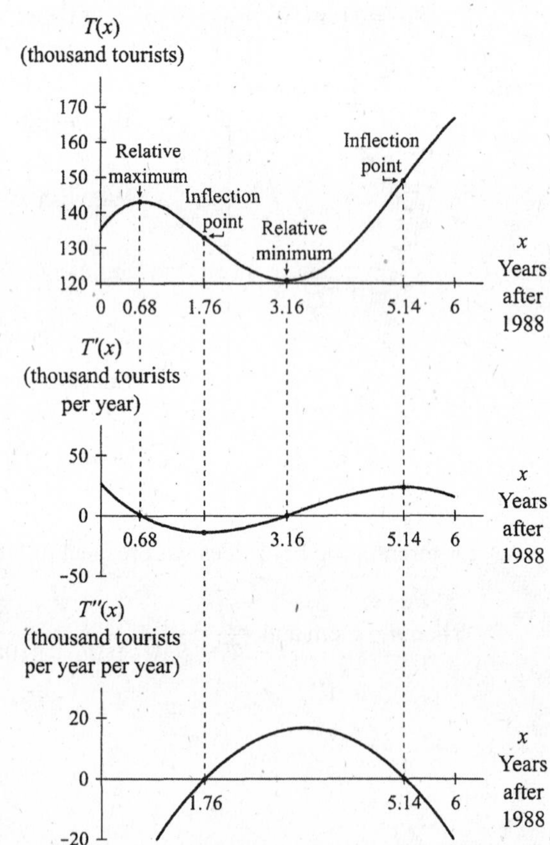

d. To determine the absolute maximum and minimum, we compare the outputs of the relative extrema with the outputs at the endpoints $x = 0$ and $x = 6$. The number of tourists was greatest in 1994 at 166.8 thousand tourists. The number was least in 1991 at 120.9 thousand.

e. To determine the greatest and least slopes, we compare the slopes at the inflection points with the slopes at the endpoints $x = 0$ and $x = 6$. The number of tourists was increasing most rapidly in 1993 at a rate of 23.1 thousand tourists per year. The number of tourists was decreasing most rapidly in 1990 at a rate of 13.3 thousand tourists per year.

2. a. (4.5 thousand people per year)$\left(\frac{1}{4}\text{ year}\right)$ = 1.125 thousand people

b. $202 + \frac{1}{2}(4.5) = 204.25$ thousand people

3. *Step 1:* Output quantity to be minimized: cost

Input quantities: distances x and y

Step 2: See Figure 5.27 in the Chapter 5 Review Test.

Step 3: Cost = $27x + 143y$ dollars for distances of x feet and y feet.

Step 4: Convert the distances in miles in the figure to distances in feet using the fact the 1 mile = 5280 feet: 3.2 miles = 16,896 feet and 1.6 miles = 8448 feet. Using the Pythagorean Theorem, we know that $8448^2 + (16{,}890 - x)^2 = y^2$. Solving for positive y yields $y = \sqrt{8448^2 + (16{,}896 - x)^2}$.

Substituting this expression for y into the equation in Step 3 gives

$C(x) = 27x + 143\sqrt{8448^2 + (16{,}896 - x)^2}$ dollars where x is the distance the pipe is run on the ground.

Step 5: Input interval: $0 < x < 16{,}890$

Step 6: The derivative of the cost function is

$$\frac{dC}{dx} = 27 + 143\tfrac{1}{2}[8448^2 + (16{,}896 - x)^2]^{-1/2}2(16{,}896 - x)(-1)$$

$$= 27 - \frac{143(16{,}896 - x)}{\sqrt{8448^2 + (16{,}896 - x)^2}}$$

Setting this equal to zero and solving for x between 0 and 16,890 gives $x = 15{,}271.7$ feet. Dividing this answer by 5280 feet per mile, we have an optimal distance of $x \approx 2{,}89$ miles

Step 7: Substituting the value of x in feet into the cost equation gives a cost of $C(15{,}271.7) \approx 1{,}642{,}527$.

4. The derivative graph lying above the axis to the left of zero and below the axis to the right of zero indicates that the graph of h increases to the left of zero and decreases to the right of zero. Thus a relative maximum occurs at $x = 0$.

The derivative graph indicates a maximum slope of h between $x = a$ and $x = 0$ and a minimum slope of between $x = 0$ and $x = b$. These points of extreme slope are inflection points on the graph of h.

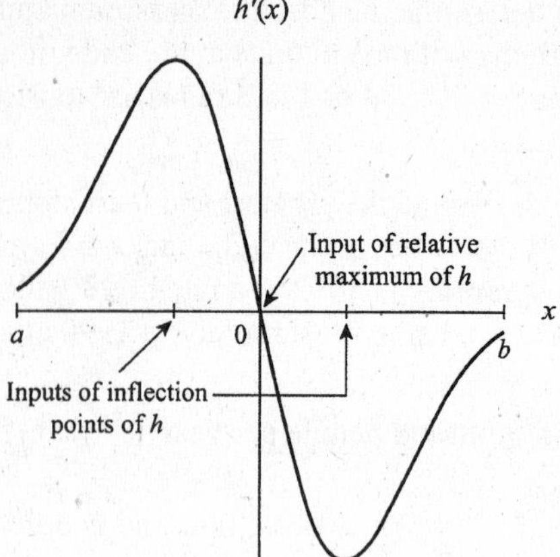

5. Treating w as a constant and differentiating the function S with respect to time yields $\frac{dS}{dt} = 0.000013w(2v)\frac{dv}{dt}$. We have $w = 4000$ pounds, $v = 60$ mph, and $\frac{dv}{dt} = -5$ mph per second. Substituting these values into the related rates equation gives
$\frac{dS}{dt} = 0.000013(4000)(2 \cdot 60)(-5) = -31.2$ feet per second.
The length of the skid marks is decreasing by 31.2 feet per second.

Chapter 5
Accumulating Change: Limits of Sums and the Definite Integral

Section 5.1 Results of Change and Area Approximations

1. a. The heights are in thousands of bacteria per hour.

b. The widths are in hours.

c. Because the area of each rectangle is a product of thousands of bacteria per hour and hours, its units are thousands of bacteria.

d. The area has the same units as the areas of the rectangles, that is, thousands of bacteria.

e. The accumulated change is a number of bacteria. Its units will be thousands of bacteria.

3. a. The area would represent how much farther a car going 60 mph will travel before stopping than a car going 40 mph.

b. i. The heights are in feet per mile per hour, and the widths are in miles per hour.

ii. Because the area of each rectangle is a product of feet per mph and mph, each rectangle area (and hence the total area) is in feet.

5. a. (thousand people per year)(years) = thousand people

b, c. Thousand people

7. a. This is the change in the number of organisms when the temperature increases from 25°C to 35°C.

b. Assuming the graph of A does not cross the c-axis between $c = 30$ and $c = 40$, this is the change in the number of organisms when the temperature increases from 30°C to 40°C.

9. a, b. Profit is increasing when the rate-of-change function is positive: between 0 and 300 boxes, and between 400 and 600 boxes.

c. N/A

d. Profit reaches a relative maximum when the rate-of-change graph passes from positive to negative: at 300 boxes.

e. Profit reaches a relative minimum when the rate-of-change graph passes from negative to positive: at 400 boxes.

f. Profit is decreasing most rapidly when the rate-of-change graph reaches a minimum: at approximately 350 boxes.

g. (dollars per box)(boxes) = dollars

h. Note that p' is the function whose graph is shown in the text. Because $p'(b) < 0$ between $b = 300$ and $b = 400$, $\int_{300}^{400} p'(b)db < 0$. Because $p'(b) > 0$ between $b = 100$ and $b = 200$, $\int_{100}^{200} p'(b)db > 0$. Therefore, $\int_{300}^{400} p'(b)db$ is less than $\int_{100}^{200} p'(b)db$.

11. a. On the horizontal axis, mark integer values of x between 0 and 8. Construct a rectangle with width from $x = 0$ to $x = 1$ and height $f(1)$. Because the width of the rectangle is 1, the area will be the same as the height. Repeat the rectangle constructions between each pair of consecutive integer input values. Note that for the rectangles that lie below the horizontal axis, the heights are the absolute values of the function values. Also note that the fourth rectangle has height 0.

Sum the areas (heights) of the rectangles to obtain the area estimate.

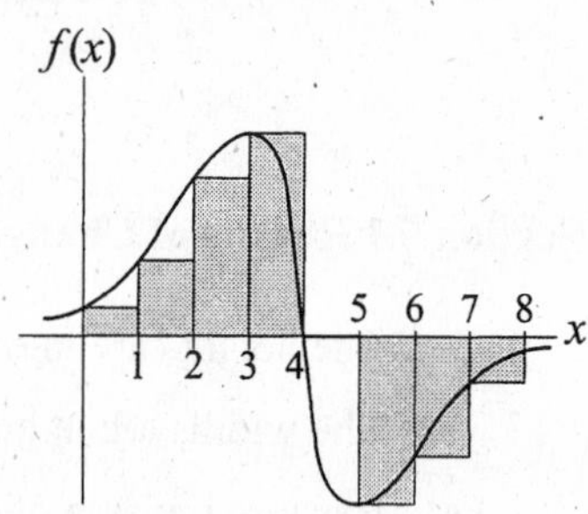

b. Repeat part *a*, except the height of each rectangle is determined by the function value corresponding to the left side of the interval. In the case of the first rectangle, the height is $f(0)$. When we use left rectangles, the fifth rectangle has height 0.

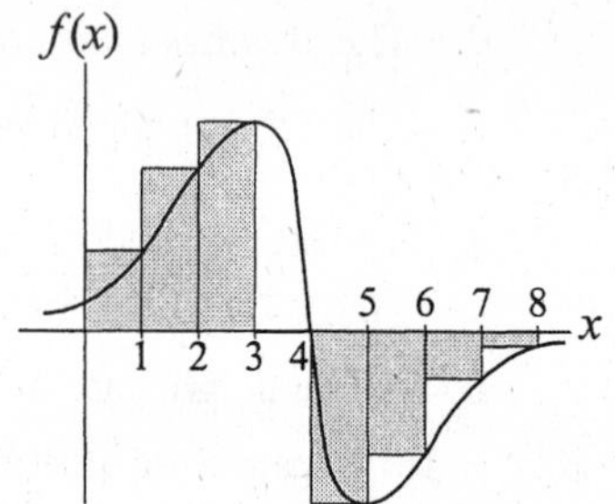

13. Divide the interval from a to b into four equal subintervals. Determine the midpoint of each subinterval and substitute the midpoints into the function to find the heights of the rectangles. Determine the area of each rectangle by multiplying the height by the width of the subintervals. Add the four areas to obtain the midpoint-rectangle estimate.

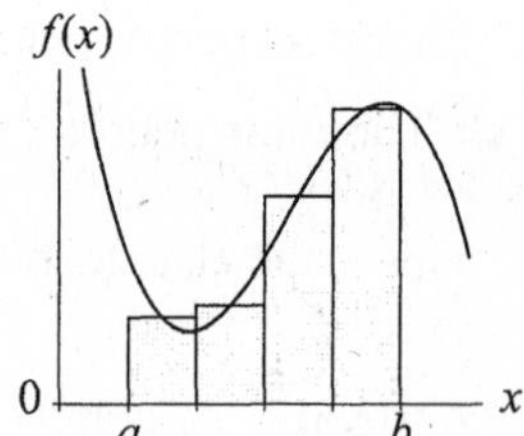

15. *One possible solution:*

a. Each rectangle has a width of 25 years.

Interval	Midpoint height (billion barrels per year)	Midpoint-rectangle area (25 years)(height)→ (billions of barrels)
1900–1925	0	0.0
1925–1950	3	75.0
1950–1975	9	225.0
1975–2000	33	825.0
2000–2025	33	825.0
2025–2050	9	225.0
2050–2075	3	75.0
2075–2100	1	25.0
Total area of rectangles ≈ 2275 billion barrels Total oil production ≈ 2275 billion barrels		

b. Each rectangle has a width of 25 years.

Interval	Midpoint height (billion barrels per year)	Midpoint-rectangle area (25 years)(height)→ (billions of barrels)
1900–1925	0	0.0
1925–1950	3	75.0
1950–1975	9	225.0
1975–2000	24	825.0
2000–2025	13	825.0
2025–2050	4	225.0
2050–2075	1	75.0
2075–2100	0	25.0
	Total area of rectangles ≈ 1625 billion barrels	
	Total oil production ≈ 1625 billion barrels	

c. The graph *A* estimate is 175 billion barrels above the journal's estimate. The graph *B* estimate is 275 billion barrels above the journal's estimate.

17. a.

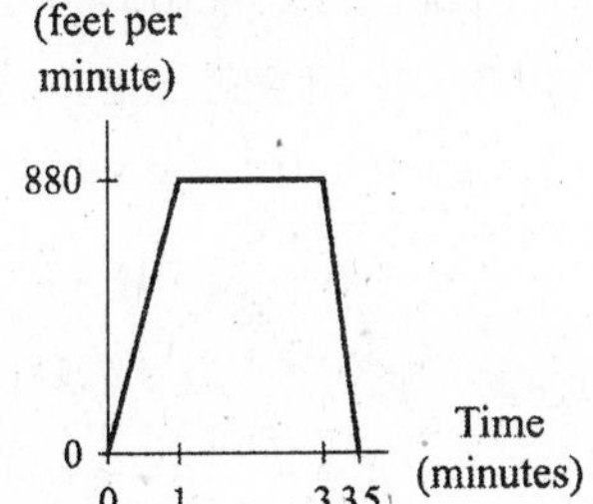

b. The region can be divided into two triangles and one rectangle. The area is calculated as

Area of triangle + area of rectangle + area of triangle

$= \frac{1}{2}(\text{base})(\text{height}) + (\text{length})(\text{width}) + \frac{1}{2}(\text{base})(\text{height})$

$= \frac{1}{2}(1 \text{ min})(880 \text{ feet/min}) + (2 \text{ min})(880 \text{ feet/min}) + \frac{1}{2}(0.5 \text{ min})(880 \text{ feet/min})$

$= 440 \text{ feet} + 1760 \text{ feet} + 220 \text{ feet} = 2420 \text{ feet}$

c. The area in part *b* is the distance the robot traveled. The robot traveled 2420 feet during the $3\frac{1}{2}$-minute experiment.

19. a.

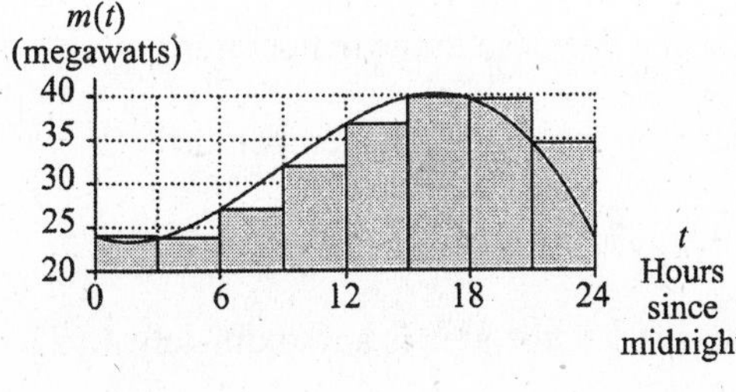

Interval	Height (megawatts)	Area = (3 hours)(height) (megawatt hours)
0 to 3	24	72.0
3 to 6	23.5	70.5
6 to 9	27.5	82.5
9 to 12	32	96.0
12 to 15	37	111.0
15 to 18	40	120.0
18 to 21	39.5	118.5
21 to 24	34.5	103.5
	Total area ≈ 774 megawatt hours	

b.

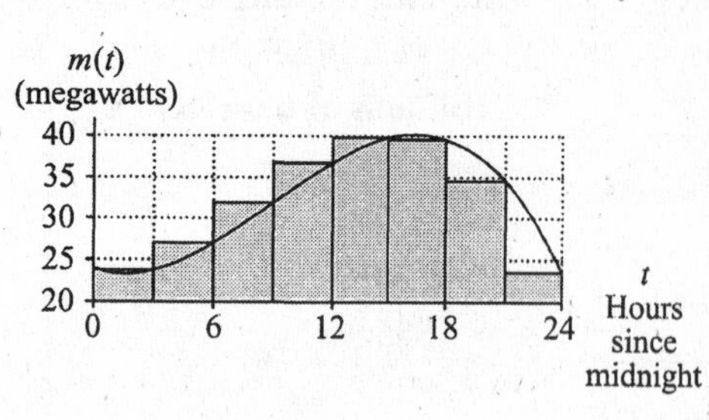

Interval	Height (megawatts)	Area = (3 hours)(height) (megawatt hours)
0 to 3	23.5	70.5
3 to 6	27.5	82.5
6 to 9	32	96.0
9 to 12	37	111.0
12 to 15	40	120.0
15 to 18	39.5	118.5
18 to 21	34.5	103.5
21 to 24	23	69.0
		Total area ≈ 771 megawatt hours

c. The exact area to three decimal places is 774.426, so both estimates are very close.

21. a.

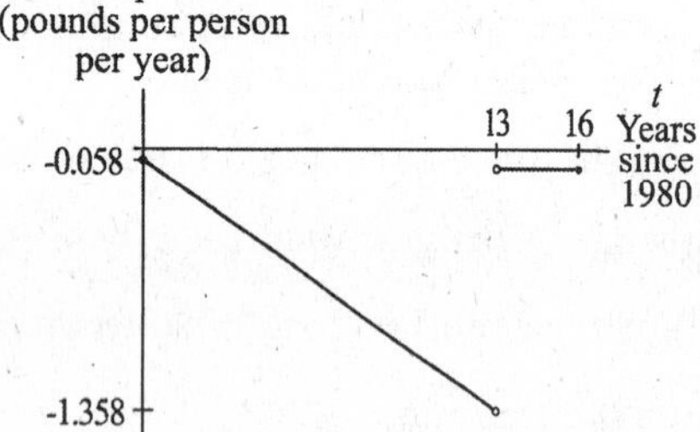

b. The area is the sum of the area of a trapezoid with width 13 and bases $|c(0)| = 0.058$ and $|-0.10(13) - 0.058| = 1.358$ and the area of a rectangle with height 0.1 and width 3:

$$\text{Area} = \left(\frac{0.058 + 1.358}{2}\right)13 + (0.1)(3)$$
$$= 9.504$$

c. Per capita consumption of cottage cheese decreased by approximately 9.5 pounds per person per year between 1980 and 1999.

d. We cannot determine the per capita consumption of cottage cheese in 1996 unless we know the value in 1980 (or some other year between 1980 and 1996). If we knew the 1980 value, we would subtract 9.5 to obtain the 1996 value. Likewise for 1999.

23. a. Solving for h in the equation $T(h) = 0$ gives $A \approx 0.907$ hour.

b. The portion below the horizontal axis represents a decrease in the temperature.

c. The portion above the horizontal axis represents an increase in the temperature.

d. Using three right rectangles from $h = 0$ to $h = 0.907$, the area is approximately 2.775.

e. Using three right rectangles from $h = 0.907$ to $h = 1.5$, the area is approximately 4.625.

f.

Number of rectangles	Approximation of area
10	-4.268
20	-4.269
40	-4.269
80	-4.270
160	-4.270
	Trend ≈ -4.270

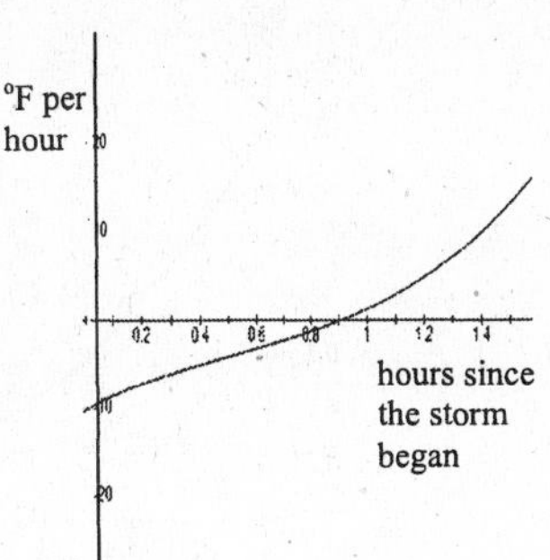

$\int_0^A T(h)dh \approx -4.270$. Between when the storm began and approximately 0.9 hours (53 minutes) later, the temperature dropped by approximately 4.27°F.

g.

Number of rectangles	Approximation of area
10	3.247
20	3.250
40	3.250
80	3.250
	Trend ≈ 3.250

$\int_A^{1.5} T(h)dh \approx 3.250$. Between 0.9 hours after the storm began and 1.5 hours after the storm began, the temperature rose by approximately 3.25°F.

e. $\int_0^{1.5} T(h)dh \approx 3.25 - 4.27 = -1.02$ °F. To determine the temperature when $h = 1.5$, we need to know the temperature at some other time (such as when the storm began or at $h = A$). The temperature was lower by approximately 1 degree.

25. a.

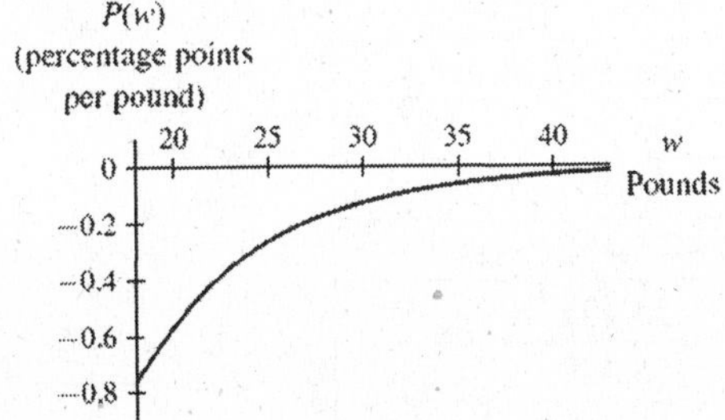

b. The fact that the graph lies below the x-axis represents that the percentage decreased between 18 and 43 pounds.

c. Using 5 midpoint rectangles, the area is approximately 4.818. The percentage decreased by approximately 4.8 percentage points when the mother's weight gain increases from 18 pounds to 43 pounds.

27. a.

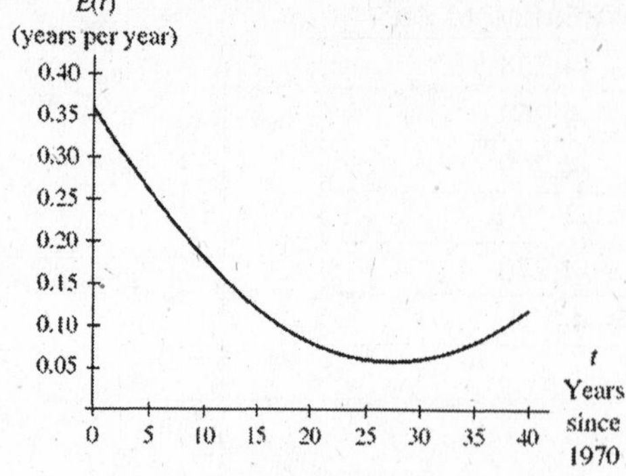

b. The point that corresponds to the life expectancy growing the least rapidly is the minimum of the graph of the rate function. This point occurs at (27.5, 0.0575)

c. The width of each rectangle is $\frac{40-0}{4} = 10$ years.

Interval (years)	Midpoint t (years)	Height $E(t)$ (years of life expectancy per year)	Area (10 years)$E(t)$ → (years of life expectancy)
0 to 10	5	0.26	2.6
10 to 20	15	0.12	1.2
20 to 30	25	0.06	0.6
30 to 40	35	0.08	0.8
		Total midpoint area ≈	**5.2 years**

From 1970 to 2010, life expectancy for women is expected to have increased by approximately 5.2 years.

d. From 2000 to 2010, life expectancy for women is expected to have increased by approximately 0.8 years. In order to determine the projected life expectancy for women in 2010 given the expected increase from 2000 to 2010, it is necessary to know the life expectancy of women in 2000.

29. a.

Number of rectangles	Approximation of area
10	55.307
20	55.407
40	55.432
80	55.438
160	55.439
320	55.440
640	55.440
1280	55.440
Trend ≈ 55.440	

Between 1990 and 2001, factory sales of electronics increased by approximately $55.44 billion.

b. $\int_0^{11} s(x)dx \approx 55.44$

c. In 2001, factory sales were approximately \$43.0 billion + \$55.44 billion = \$98.4 billion.

31. a. $S(t) = \dfrac{695.606}{1 + 0.081e^{0.495438t}}$ thousand DVDs per month t months after release

b.

Number of rectangles	Approximation of area
5	3617.9
10	3623.6
15	3625.4

The estimates are approximately 3,618,000 DVDs, 3,624,000 DVDs, and 3,625,000 DVDs.

c. The sum of the sales figures (iii) will be most accurate because it will give the exact total. (Note that the text does not give all of the data needed to perform this calculation.)

33. a.

Number of rectangles	Approximation of area
5	9.8577
10	10.0954
20	10.0987

The estimates are approximately 9.8577 million labor hours, 10.0954 million labor hours, and 10.0987 million labor hours.

b. Because the activity specifies that the function evaluated at the ends of weeks (integers) gives the exact number of labor hours needed, simply adding the function at integer values will give an exact answer for the total number of worker hours. This is the sum of right rectangles (i). Of course, this exact answer doesn't give the exact area beneath the curve.

35. *One possible answer:* The definite integral gives the accumulated change. They are both related to the area trapped between the rate curve and the horizontal axis. If the area is above the horizontal axis, the definite integral is positive and the accumulated change represents an increase. If the area is below the horizontal axis, the definite integral is negative and the accumulated change represents a decrease.

Section 5.2 Limit of Sums, Accumulated Change, and The Definite Integral

1. a.

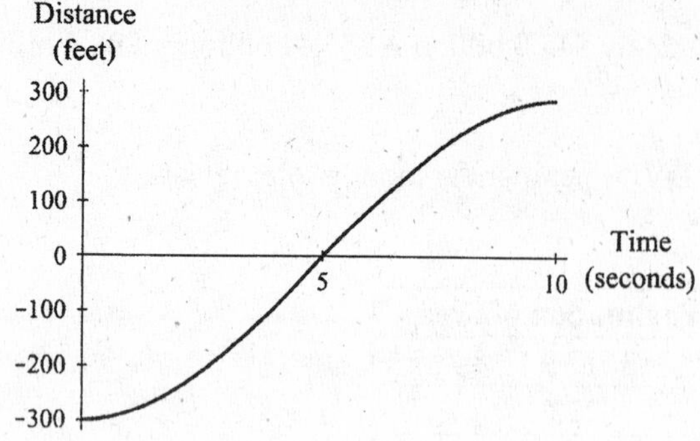

b. $D(x)=\int_5^x f(t)dt$

c. The accumulation function gives the distance traveled between 5 seconds and x seconds. For times before 5 seconds, the accumulation function is the negative of the distance traveled, because we are looking back in time.

3. a. The area of the region between days 0 and 18 represents how much the price of the technology stock declined ($15.40 per share) during the first 18 trading days of 2003.

b. The area of the region between days 18 and 47 represents how much the price of the technology stock rose ($55.80) between days 18 and 47.

c. The price on day 47 was $40.40 more than the price on day 0.

d. The price was $11.10 less on day 55 than it was on day 47.

e.

x	0	8	18	35	47	55
$\int_0^x r(t)dt$	0	-7.1	$-(7.1+8.3)=$ -15.4	$-15.4+30.4=$ 15.0	$15+25.4$ $=40.4$	$40.4-11.1=$ 29.3

f.

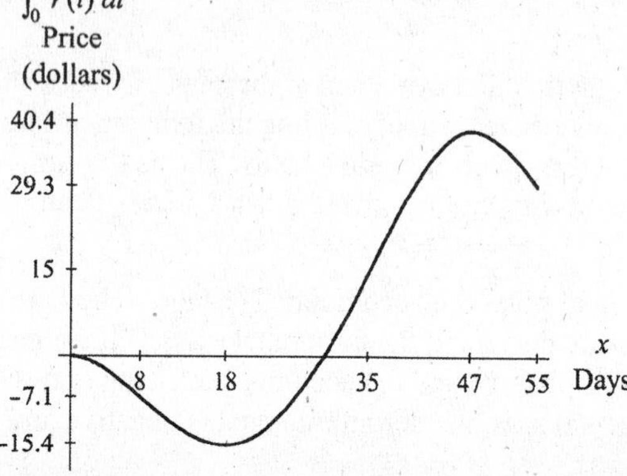

g. On day 55 the stock price was $127 + $29.30 = $156.30.

5. a. Because the rate-of-change graph is always positive, lying above the horizontal axis, the number of subscribers never declined during the first year.

b. The peak corresponds to the time when the number of subscribers was increasing most rapidly, that is, when the rate of change was greatest.

c. The accumulation function represents the change in the number of subscribers between day 140 and day t, in other words, how many new subscribers are added t days after the end of the twentieth week.

d. Each box has width 4 weeks or 28 days and height 10 subscribers per day. Multiplying these values gives an area of (28 days)(10 subscribers per day) = 280 subscribers.

e. Estimates will vary. Multiply the number of boxes by 280 subscribers per box to obtain the area estimate. Possible estimates are as follows:

Week	t (days)	Number of boxes	Area = $\int_0^t n(x)dx$	Week	t (days)	Number of boxes	Area = $\int_0^t n(x)dx$
4	28	1.25	350	28	196	31.5	8820
8	56	3.3	924	36	252	37.25	10,430
12	84	7	1960	44	308	39	10,920
16	112	12.5	3500	52	364	39.5	11,060
20	140	19.25	5390				

f. $\int_{140}^{t} n(x)\,dx$

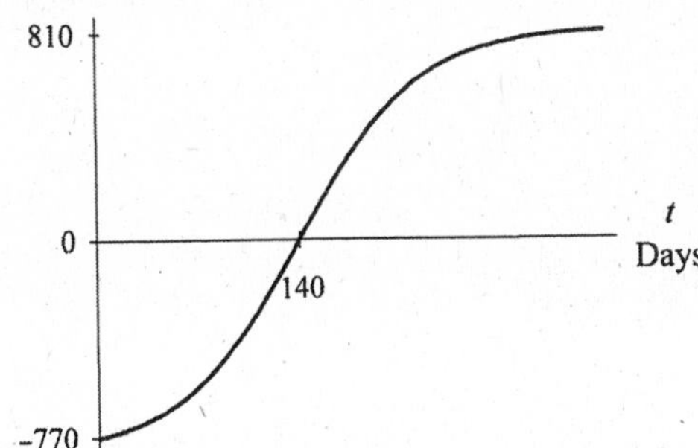

7.

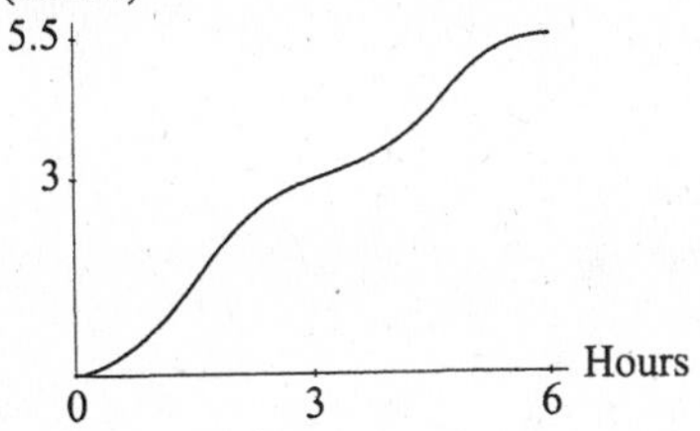

9. a. $\int_A^x f(t)\,dt$

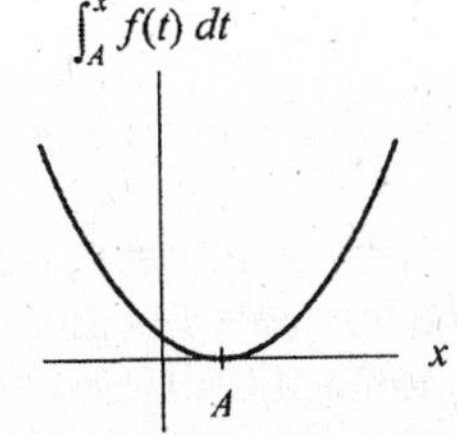

11. **a.** $\int_0^x f(t)\,dt$

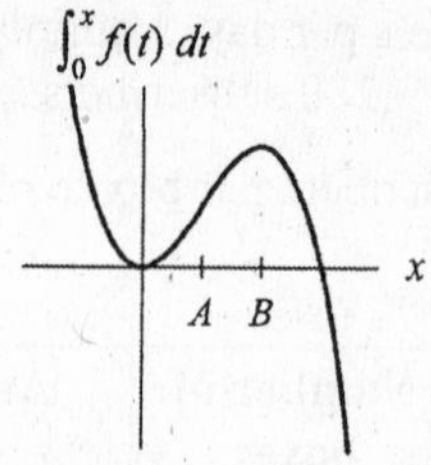

b. $\int_A^x f(t)\,dt$

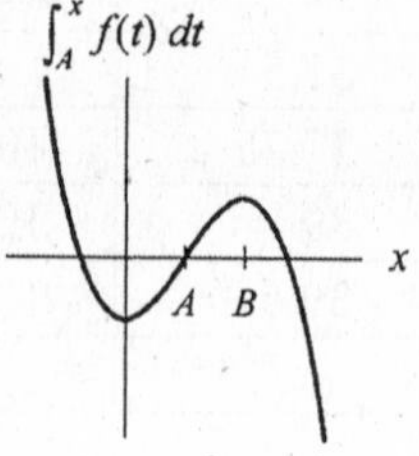

c. $\int_B^x f(t)\,dt$

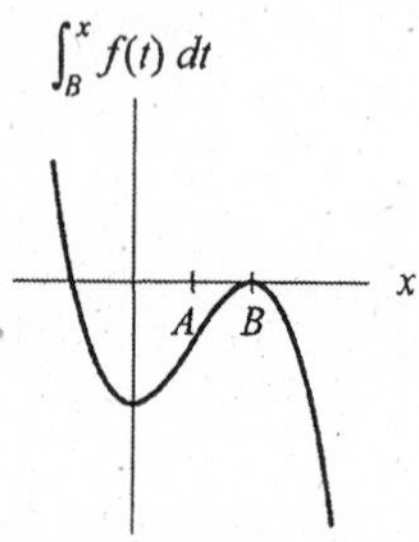

13. $\int_0^x f(t)\,dt$

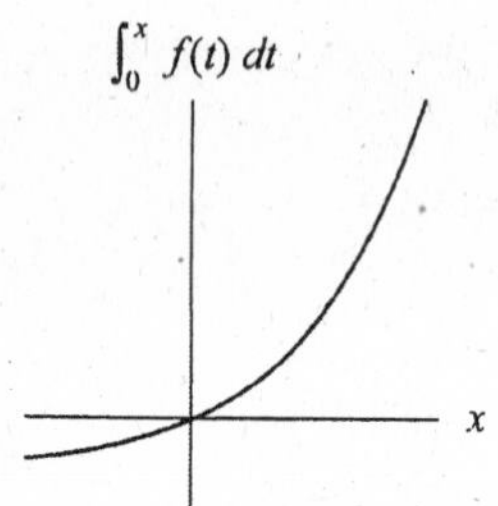

15. Derivative graph: b
Accumulation graph: f

17. Derivative graph: f
Accumulation graph: e

19. Because the table values indicate that a graph of f decreases between $t = 0$ and $t = 4$, we expect the derivative values to be non-positive between $t = 0$ and $t = 4$. The left table fits this description and is thus the derivative table. The accumulation function is the table on the right, with values increasing between $t = 0$ and $t = 2$ as the area accumulates above the x-axis and then decreasing between $t = 2$ and $t = 4$ indicating the negative accumulation.

21. a. $\dfrac{\text{million dollars of revenue}}{\text{thousand advertising dollars}}$

b.

Input: m thousands of advertising dollars

Rule R

Output: $R(m)$ millions of dollars of revenue

c. When m thousand dollars are being spent on advertising, the annual revenue is $R(m)$ million dollars.

23. a. 

b. 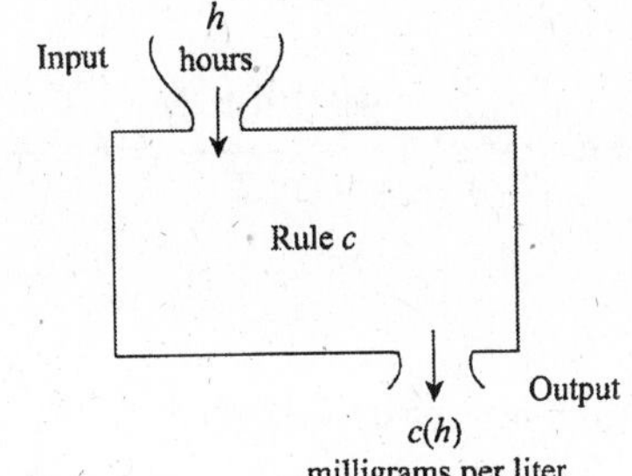

c. The concentration of a drug in the bloodstream is $c(h)$ milligrams per liter h hours after the drug is given.

25. *One possible answer:* The fact that the rate function is negative indicates that the accumulation is negative and thus the accumulation graph decreases. The fact that the function is increasing does not affect whether or not the accumulation graph increases or decreases.

Section 5.3 Accumulation Functions

1. a. ii **b.** i **c.** iii

3. a. iii **b.** i **c.** ii

5. *One possible answer:* Using the function $f(t) = 3t$ and the constant $a = 5$, we illustrate the Fundamental Theorem of Calculus as follows:

$$\frac{d}{dx}\int_5^x 3t\,dt = \frac{d}{dx}\left(\frac{3t^2}{2}\Bigg|_5^x\right) = \frac{d}{dx}\left(\frac{3x^2}{2} - \frac{75}{2}\right) = 3x$$

7. *One possible answer:* The Fundamental Theorem of Calculus tells us that for a continuous function, the derivative of an antiderivative of the function gives back the function itself.

9. $\int 19.4(1.07^x)dx = 19.4\int 1.07^x dx$

$$= 19.4\left(\frac{1}{\ln 1.07}\right)1.07^x + C$$

$$= \frac{19.4(1.07^x)}{\ln 1.07} + C$$

Check: $\frac{d}{dx}\left(\frac{19.4(1.07^x)}{\ln 1.07} + C\right) = \frac{19.4(1.07^x)(\ln 1.07)}{\ln 1.07} = 19.4(1.07^x)$

11. $\int\left[6e^x + 4\left(2^x\right)\right]dx = 6\int e^x dx + 4\int 2^x dx$

$$= 6e^x + 4\left(\frac{1}{\ln 2}\right)2^x + C$$

$$= 6e^x + \frac{4\left(2^x\right)}{\ln 2} + C$$

Check: $\frac{d}{dx}\left(6e^x + \frac{4\left(2^x\right)}{\ln 2} + C\right) = 6e^x + \frac{4\left(2^x\right)(\ln 2)}{\ln 2} = 6e^x + 4\left(2^x\right)$

13. $\int\left(10^x + 4\sqrt{x} + 8\right)dx = \int 10^x dx + 4\int x^{1/2}dx + \int 8dx$

$$= \frac{10^x}{\ln 10} + \frac{4}{3/2}x^{3/2} + 8x + C$$

$$= \frac{10^x}{\ln 10} + \frac{8}{3}x^{3/2} + 8x + C$$

Check: $\frac{d}{dx}\left(\frac{10^x}{\ln 10} + \frac{8}{3}x^{3/2} + 8x + C\right) = \frac{10^x(\ln 10)}{\ln 10} + \frac{8}{3}\left(\frac{3}{2}x^{1/2}\right) + 8 = 10^x + 4\sqrt{x} + 8$

15. $S(m) = \int s(m)dm$

$$= \int 600(0.93^m)dm$$

$$= 600\int 0.93^m dm$$

$$= 600\left(\frac{1}{\ln 0.93}\right)0.93^m + C$$

$$= \frac{600(0.93^m)}{\ln 0.93} + C \text{ CDs}$$

m months after the beginning of the year

17. $C(x) = \int c(x)dx$

$$= \int\left(\frac{0.8}{x} + 0.38(0.01^x)\right)dx$$

$$= 0.8\int\frac{1}{x}dx + 0.38\int 0.01^x\,dx$$

$$= 0.8\ \ln|x| + \frac{0.38(0.01^x)}{\ln 0.01} + K$$ dollars per unit when x units are produced

Note: If the production is presumed positive, the absolute value symbol is not necessary.

19. Find the general antiderivative:

$$F(x) = \int f(x)dx$$

$$= \int\left(t^2 + 2t\right)dt$$

$$= \int t^2 dt + 2\int t\ dt$$

$$= \frac{t^3}{3} + 2\left(\frac{t^2}{2}\right) + C$$

$$= \frac{t^3}{3} + t^2 + C$$

Solve for C: $F(12) = 700$

$$\frac{12^3}{3} + 12^2 + C = 700$$

$$720 + C = 700$$

$$C = -20$$

The specific antiderivative is $F(t) = \frac{1}{3}t^3 + t^2 - 20$

21. Find the general antiderivative:

$$F(z) = \int f(z)dz$$

$$= \int\left(z^{-2} + e^z\right)dz$$

$$= \frac{z^{-1}}{-1} + e^z + C$$

$$= \frac{-1}{z} + e^z + C$$

Solve for C: $F(2) = 1$

$$\tfrac{-1}{2} + e^2 + C = 1$$

$$C = \tfrac{3}{2} - e^2$$

The specific antiderivative is $F(z) = \frac{-1}{z} + e^z + \left(\tfrac{3}{2} - e^2\right)$

23. Find the general antiderivative:

$$w(t) = \int \frac{7.37}{t} dt = 7.37 \ln t + C$$

(Assume t is positive.)
Solve for C. At 9 weeks, $t + 2 = 9$, so $t = 7$ and $w(7)=26$.

$$7.37 \ln 7 + C = 26$$
$$C = 26 - 7.37 \ln 7 \approx 11.659$$

The specific antiderivative is $w(t) = 7.37 \ln t + 26 - 7.37 \ln 7$
$\approx 7.37 \ln t + 11.659$ grams after $(t + 2)$ weeks

This specific antiderivative is the formula for the accumulation function of W passing through the point (7, 26).

25. a. Find the general antiderivative: $G(t) = \int \left[(1.667 \cdot 10^{-4})t^2 - 0.02t - 0.10\right] dt$

$$= \frac{1.667 \cdot 10^{-4}}{3} t^3 - \frac{0.02}{2} t^2 - 0.10t + C$$

Solve for C using the fact that $G(70) = 94.8$:

$$\frac{1.667 \cdot 10^{-4}}{3}(70^3) - \frac{0.02}{2}(70^2) - 0.10(70) + C = 94.8$$
$$C \approx 131.741$$

The specific antiderivative is $G(t) = \frac{1.667 \cdot 10^{-4}}{3} t^3 - \frac{0.02}{2} t^2 - 0.10t + 131.741$ males per 100 females t years after 1900.

b. This specific antiderivative is the formula for the accumulation function of g passing through (70, 94.8).

27. a. Find the general antiderivative: $v(t) = \int a(t)dt = \int (-32)dt = -32t + C$

Solve for C:
$$v(0) = 0$$
$$-32(0) + C = 0$$
$$C = 0$$

The specific antiderivative is $v(t) = -32t$ ft/sec t seconds after the penny is dropped.

b. Find the general antiderivative:0

$$s(t) = \int v(t)dt = \int -32t\,dt = -32 \int t\,dt = -32\frac{t^2}{2} + K = -16t^2 + K$$

Solve for C:
$$s(0) = 540$$
$$-16(0)^2 + K = 540$$
$$K = 540$$

The specific antiderivative is $S(t) = -16t^2 + 540$ ft above ground level t seconds after the penny is dropped.

c.

$$s(t) = 0$$
$$-16t^2 + 540 = 0$$
$$t^2 = 33.75$$
$$t \approx \pm 5.809$$

The penny will hit the ground approximately 5.8 seconds after it is dropped.

29. a. $a(t) = -32$ ft/sec^2

$v(t) = \int a(t)dt = -32t + C$

Because $v(0) = 0$, $v(t) = -32t$ feet per second after t seconds.

$s(t) = \int v(t)dt = -16t^2 + C$

Because $s(0) = 66$, $s(t) = -16t^2 + 66$ feet after t seconds.

Solving $s(t) = 0$ gives $t \approx 2.031$ sec.

The impact velocity is $v(2.031) \approx -64.99$ ft/sec.

$$\left(\frac{-64.99 \text{ ft}}{1 \text{ sec}}\right)\left(\frac{3600 \text{ sec}}{1 \text{ hour}}\right)\left(\frac{1 \text{ mile}}{5280 \text{ ft}}\right) \approx \frac{-44.31 \text{ miles}}{1 \text{ hour}}$$

The impact velocity is –44.31 mph.

b. Air resistance probably accounts for the difference.

31. a. Find the general antiderivative: $N(x) = \int n(x)dx = \int\left(\frac{593}{x} + 138\right)dx$

$$= 593\int\frac{1}{x}dx + \int 138dx = 593 \ln|x| + 138x + C$$

Solve for C:

$$N(5) = 896$$
$$593\ln 5 + 138(5) + C = 896$$
$$C \approx -748.397$$

The specific antiderivative is $N(x) = 593 \ln x + 138x - 748.397$ employees x years after 1996. *Note:* Because x is positive for the years when the model is valid, the absolute value symbol is not needed.

b. The function in part *a* applies from 1997 ($x = 1$) through 2002 ($x = 6$).

c. There are two ways to estimate the number of employees the company hired. If we consider the function to be continuous with discrete interpretation, then the number can be calculated from the function n by summing the yearly totals:

$n(1) + n(2) + n(3) + n(4) + n(5) + n(6) \approx 2281$

We can also estimate the total number of employees hired between 1997 and 2002 as $\int_1^6 n(x)dx = N(6) - N(1) \approx 1753$. This estimate treats the function as continuous without restriction and is probably less accurate than the first estimate.

If any employees were fired or quit between 1997 and 2002, an estimate of the number of employees hired would not represent the number of employees at the end of 2002.

Section 5.4 The Fundamental Theorem

1. c **3.** c **5.** b **7.** a

9. a.

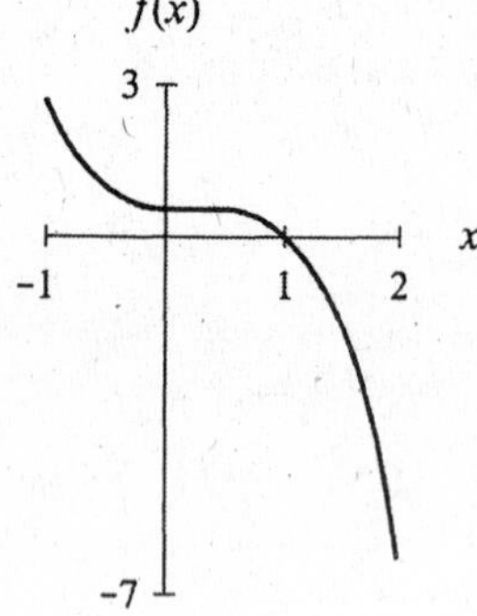

b. Begin by finding the general antiderivative of *f*:

$$\int f(x)dx = \int(-1.3x^3 + 0.93x^2 + 0.49)dx$$

$$= \frac{-1.3x^4}{4} + \frac{0.93x^3}{3} + 0.49x + C$$

$$= -0.325x^4 + 0.31x^3 + 0.49x + C$$

By solving $f(x) = 0$, we find that the graph crosses the horizontal axis at $x \approx 1.0544$.

$$\text{Area} = \int_{-1}^{1.0544} f(x)dx - \int_{1.0544}^{2} f(x)dx$$

$$= (-0.325x^4 + 0.31x^3 + 0.49x)\Big|_{-1}^{0.544} - (0.325x^4 + 0.31x^3 + 0.49x)\Big|_{1.0544}^{2}$$

$$\approx 0.478 - (-1.125) - [-1.74 - 0.478]$$

$$\approx 3.822$$

Because the graph of *f* crosses the *x*-axis between $x = -1$ and $x = 2$, $\int_{-1}^{2} f(x)dx$ is not the area found in part *b*.

c. $\int_{-1}^{2} f(x)dx = -0.615$

11. a.

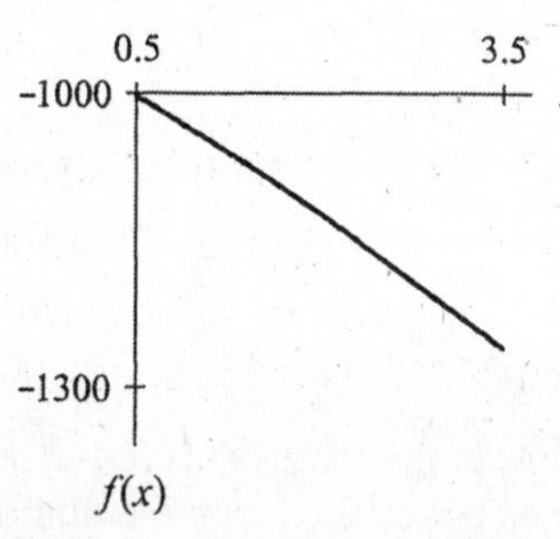

b. $$\text{Area} = -\int_{0.5}^{3.5} f(x)dx$$

$$= -\int_{0.5}^{3.5} -965.27(1.079^x)dx$$

$$= \int_{0.5}^{3.5} 965.27(1.079^x)dx$$

$$= \frac{965.27(1.079^x)}{\ln 1.079}\Bigg|_{0.5}^{3.5}$$

$$\approx 16{,}565.788 - 13{,}187.054$$

$$\approx 3378.735$$

Because the graph of f lies below the *x*-axis, $\int_{0.5}^{3.5} f(x)dx$ is the signed area (the negative of the area) calculated in part *b*.

c. $\int_{0.5}^{3.5} f(x)dx \approx -3378.735$

13. $\int_{5}^{15} P(x)dx \approx 2305.357$ Between 1985 and 1995, the number of international calls billed in the United States increased by 2305.4 million calls.

15. $\int_{0}^{5} r(x)dx = \int_{0}^{5}\left(9.907x^2 - 40.769x + 58.492\right)dx$

$$\approx \left(3.302x^3 - 20.385x^2 + 58.492x\right)\Big|_0^5$$

$$\approx 195.639 - 0$$

$$= 195.639$$

The corporation's revenue increased by \$195.6 million between 1987 and 1992.

17. a. $\int_{0}^{70} s(t)dt = \int_{0}^{70}(0.00241t + 0.02905)dt$

$$= \left(0.001205t^2 + 0.02905t\right)\Big|_0^{70}$$

$$= 7.938 - 0$$

$$= 7.938$$

In the 70 days after April 1, the snow pack increased by 7.938 equivalent cm of water.

b. $\int_{72}^{76} s(t)dt = \int_{72}^{76}\left(1.011t^2 - 147.941t + 5406.578\right)$

$$\approx \left(0.337t^3 - 73.9855t^2 + 5406.578t\right)\Big|_{72}^{76}$$

$$= 131{,}494.592 - 131{,}517.36$$

$$= -22.768$$

Between 72 and 76 days after April 1, the snow pack decreased by 22.768 equivalent centimeters of water.

c. It is not possible to find $\int_{0}^{76} s(t)dt$ because $s(t)$ is not defined between $t = 70$ and $t = 72$.

19. a. Using technology, we find that the graph crosses the horizontal axis at $h \approx 0.8955$.

$$\text{Area} \approx \int_{0}^{0.8955} T(h)dh$$

$$= \int_{0}^{0.8955}\left(9.07h^3 - 24.69h^2 + 14.87h - 0.03\right)dh$$

$$= \left(2.2675h^4 - 8.23h^3 + 7.435h^2 - 0.03h\right)\Big|_0^{0.8955}$$

$$\approx 1.483 - 0 = 1.483$$

From 8:30 a.m. to 8:54 a.m., the temperature increased by 1.48°F.

b. Area $\approx -\int_{0.8955}^{1.75} T(h)dh$

$$= -\int_{0.8955}^{1.75} \left(9.07h^3 - 24.69h^2 + 14.87h - 0.03\right)dh$$

$$= -\left(2.2675h^4 - 8.23h^3 + 7.435h^2 - 0.03h\right)\Big|_{0.8955}^{1.75}$$

$$\approx -(-0.124 - 1.483) = 1.607$$

After rising 1.48°F, the temperature decreased by 1.61°F between 8:54 a.m. to 10:15 a.m.

c. No, the highest temperature reached was 71 + 1.48 = 72.48°F.

21. a. An exponential model for the data is: The rate of change of marketed natural gas is $f(x) = 0.161(1.076186^x)$ trillion cubic feet per year x years after 1900.

b. $\int_{40}^{60} f(x)dx = \int_{40}^{60} 0.161(1.076186^x)dx = \frac{0.161(1.076186^x)}{\ln 1.076186}\Big|_{40}^{60} \approx 179.725 - 41.387 = 138.338$

From 1940 through 1960, 138.3 trillion cubic feet of natural gas was produced.

c. $\int_{40}^{60} f(x)dx$

23. a. The marginal cost for an additional CD is $C'(x) = \left(7.714 \cdot 10^{-5}\right)x^2 - 0.047x + 8.940$ dollars per CD, when x CDs are produced each hour.

b. Find the general antiderivative: $C(x) = \int C'(x)dx$

$$= \int\left[\left(7.714 \cdot 10^{-5}\right)x^2 - 0.047x + 8.94\right]dx$$

$$= \left(2.571 \cdot 10^{-5}\right)x^3 - 0.024x^2 + 8.940x + K$$

Use the fact that C(150)=750 to solve for K:

$$\left(2.571 \cdot 10^{-5}\right)(150)^3 - 0.024(150)^2 + 8.940(150) + K = 750$$

$$K \approx -143.893$$

The hourly cost model is $C(x) = \left(2.571 \cdot 10^{-3}\right)x^3 - 0.024x^2 + 8.940x - 143.893$ dollars when x CDs are produced each hour.

c. $\int_{200}^{300} C'(x) = C(x)\Big|_{200}^{300} = C(300) - C(200) \approx 1096.82 - 900.68 = \$196.14$

When production is increased from 200 to 300 CDs per hour, cost increases by approximately \$196.14.

25. a, b.

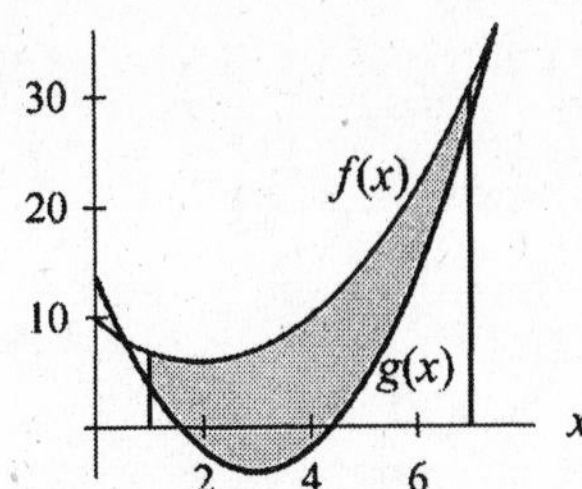

c. Area:

$$\int_a^b [f(x) - g(x)]dx$$

$$= \int_1^7 [(x^2 - 4x + 10) - (2x^2 - 12x + 14)]dx$$

$$= \int_1^7 (-x^2 + 8x - 4)dx$$

$$= \left(\frac{-x^3}{3} + 4x^2 - 4x\right)\Bigg|_1^7$$

$$= \frac{161}{3} - \left(\frac{-1}{3}\right) = 54$$

27. a, c.

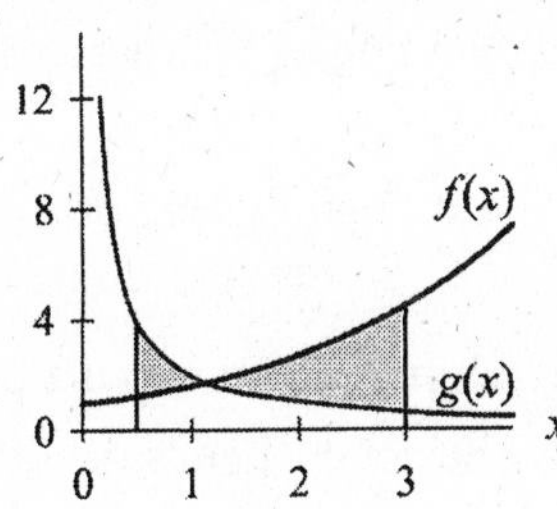

b. Using technology, we find that the curves intersect at $x \approx 1.134$.

d. Difference $= \int_{0.5}^{3} [f(x) - g(x)]dx$

$$= \int_{0.5}^{3} \left[e^{0.5x} - \frac{2}{x}\right]dx$$

$$= \left[\frac{e^{0.5x}}{0.5} - 2\ln|x|\right]\Bigg|_{0.5}^{3}$$

$$= (2e^{0.5x} - 2\ln x)\Big|_{0.5}^{3}$$

$$\approx 6.7662 - 3.9543 \approx 2.812$$

e. Area of left region $\approx \int_{0.5}^{1.134} [g(x) - f(x)]dx$

$= (2\ln x - 2e^{0.5x})\Big|_{0.5}^{1.134}$

$\approx -3.274 - (-3.954)$

$= 0.680$

Area of right region $\approx \int_{1.134}^{3} [f(x) - g(x)]dx$

$= (2e^{0.5x} - 2\ln x)\Big|_{0.5}^{3}$

$\approx 6.766 - 3.274$

$= 3.492$

Total area $\approx 0.680 + 3.492 = 4.172$

29. a. When the amount invested in capital increases from \$1500 to \$5500, profit increases by approximately \$13.29 million.

b. Area $= \int_{1.5}^{5.5} [r'(x) - c'(x)]dx = 13.29$

31. a. The population of the country grew by 3690 people in January.

b. The population declined by 9720 people between the beginning of February and the beginning of May.

c. The change in population from the beginning of January through the end of April was $3690 - 9720 = -6030$ people

d. Because the graphs intersect, the area of R_1 represents an increase in population and the area of region R_2 represents a decrease in population. The net change is the difference:

Area of R_1 − Area of R_2

The total area is

Area of R_1 + Area of R_2

33. a. Before fitting models to the data, add the point (0, 0), and convert the data from miles per hour to feet per second by multiplying each speed by $\left(\frac{5280 \text{ feet}}{1 \text{ mile}}\right)\left(\frac{1 \text{ hour}}{3600 \text{ seconds}}\right)$. For each car, the speeds in the revised data set are 0, 44, $58\frac{2}{3}$, $73\frac{1}{3}$, 88, $102\frac{2}{3}$, $117\frac{1}{3}$, 132, and $146\frac{2}{3}$ feet per second.

The speed of the Supra after t seconds can be modeled as $s(t) = -0.702t^2 + 20.278t + 2.440$ feet per second.

The speed of the Carrera after t seconds can be modeled as $c(t) = -0.643t^2 + 18.963t + 5.252$ feet per second.

b. $\int_0^{10}[s(t)-c(t)]dt \approx \int_0^{10}(-0.059t^2+1.315t-2.811)dt$

$\approx (-0.020t^3+0.657t^2-2.811t)\Big|_0^{10}$

$\approx 17.965-0=17.965$

The Supra travels approximately 17.96 feet farther.

c. $\int_5^{10}[s(t)-c(t)] = (-0.020t^3+0.657t^2-2.811t)\Big|_0^{10}$

$\approx 17.965-(-0.079)=18.044$

The Supra travels approximately 18.04 feet farther.

35. a. FedEx: $F(t)=-0.026t^3+0.198t^2+0.06t+0.317$ billion dollars per year t years after 1993

UPS: $U(t)=0.15t+0.572$ billion dollars per year t years after 1993

b. Area of region on left $\approx$ \$1.28 billion
Area of region in middle $\approx$ \$1.20 billion
Area of region on right $\approx$ \$1.59 billion

Between the beginning of 1993 and late 1996 ($t \approx 2.8$), UPS's accumulated revenue exceeded that of FedEx by approximately \$1.28 billion.

Between late 1996 and the spring of 2000 ($t \approx 6.3$), FedEx's accumulated revenue exceeded that of UPS by approximately \$1.2 billion.

Between the spring of 2000 and the end of 2001, UPS's accumulated revenue exceeded that of FedEx by approximately \$1.59 billion.

c. $\int_3^{11}[F(t)-U(t)]dt = 1.675$ billion. This value is the net amount by which FedEx's accumulated revenue exceeded that of UPS between 1993 and 2001.

37. a. Multiply the output data by 22 to convert the data to total absorption for all 22 hectares of trees. Finding a logistic model for these converted data, we have

$$f(x)=\begin{cases} 0 \text{ tons per year} & \text{when } 0 \le x < 5 \\ \dfrac{557.960}{1+91.202e^{-0.318025x}} \text{ tons per year} & \text{when } x \ge 5 \end{cases}$$

x years after 1990.

b,c.

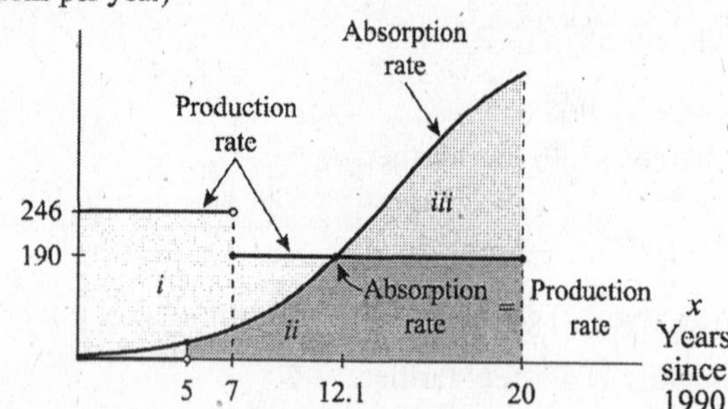

d. i. Between 1990 and 1995, the factory produced (246 tons per year)(5 years) = 1230 tons. Between 1995 and 1997, the amount the factory produced that was not absorbed by the trees is calculated as $\int_5^7 [246 - f(x)]dx \approx 414.2$ tons. To find the time when the trees began absorbing more than the factory produced, we solve $f(x) = 190$ and find $x \approx 12.1$. Between 1997 and early in 2003, the amount the factory produced that was not absorbed by the trees is calculated as $\int_7^{12.1} [190 - f(x)]dx \approx 410.7$ tons.

Summing these three totals, we have the total emissions produced by the factory and not absorbed by the trees: $1230 + 414.2 + 410.7 = 2054.9$ tons

ii. The emissions produced by the factory and absorbed by the trees is found as

$$\int_5^{12.1} f(x)dx + (190)(20 - 12.1) = 638.5 + 1498.6 = 2137.1 \text{ tons}$$

iii. The emissions absorbed by the trees from sources other than the factory is calculated as
$\int_{12.1}^{20} [f(x) - 190]dx \approx 1269.0$ tons

e. The factory produces (246 tons per year)(7 years) + (190 tons per year)(13 years) = 4192 tons, and the trees absorb $\int_5^{20} f(x)dx \approx 3406.2$ tons. This does not comply with the federal regulation.

39. *One possible answer:* Take the absolute value of the difference of the heights of each curve.

Section 5.5 Average Value and Average Rates of Change

1. a. $V(t) = -1.664t^3 + 5.867t^2 + 1.640t + 60.164$ mph t hours after 4 p.m.

b.
$$\frac{\int_0^3 V(t)dt}{3-0} \approx \frac{1}{3}\int_0^3 (-1.664t^3 + 5.867t^2 + 1.640t + 60.164)dt$$
$$\approx \frac{1}{3}(-0.416t^4 + 1.956t^3 + 0.820t^2 + 60.164t)\Big|_0^3$$
$$\approx \frac{1}{3}(206.97) \approx 68.99 \text{ mph}$$

c. $\dfrac{\int_1^3 V(t)dt}{3-1} \approx \dfrac{1}{2}(-0.416t^4 + 1.956t^3 + 0.820t^2 + 60.164t)\Big|_1^3$

$\approx \dfrac{1}{2}(206.97 - 62.52) \approx 72.23\text{mph}$

3. a.

$$R(y) = \begin{cases} \dfrac{0.719}{1+0.005e^{0.865563y}} + 0.62 \text{ dollars per minute} & \text{when } 2 \le y < 10 \\ -0.045y + 1.092 \text{ dollars per minute} & \text{when } 10 \le y \le 20 \end{cases}$$

y years after 1980

b. $\dfrac{\int_2^{10} R(y)dy}{10-2} = \dfrac{1}{8}\int_2^{10}\left(\dfrac{0.719}{1+0.005e^{0.865563y}} + 0.62\right)dy \approx \$0.99$ per minute

c. $\dfrac{\int_2^{20} R(y)dy}{20-2} = \dfrac{1}{18}\left[\int_2^{10}\left(\dfrac{0.719}{1+0.005e^{0.865563y}} + 0.62\right) + \int_{10}^{20}(-0.045y + 1.092)dy\right]$

$\approx \dfrac{1}{18}(7.925 + 4.167) \approx \$0.67$ per minute

5. a. Note that the beginning of 1990 is the end of 1989, corresponding to $t = 89$.

Let $P(t) = 7.567(1.02639^t)$.

$\dfrac{\int_{89}^{99} P(t)dt}{99-89} = \dfrac{1}{10}\left(\dfrac{7.567(1.02639^t)}{\ln 1.02639}\right)\Bigg|_{89}^{99}$

$\approx \dfrac{1}{10}(3828.97 - 2950.92)$

≈ 87.8 million people

b. Solving $P(t) \approx 87.8$ gives $t \approx 94.1$ years after the end of 1900. This corresponds to early 1995.

c. Average rate $= \dfrac{P(99) - P(89)}{99-89}$

$\approx \dfrac{99.74 - 76.86}{10} \approx 2.29$ million people per year

7. a. Because the model is linear, the coefficient of x gives the rate of change: -100.6 yearly accidents per year.

b, c.

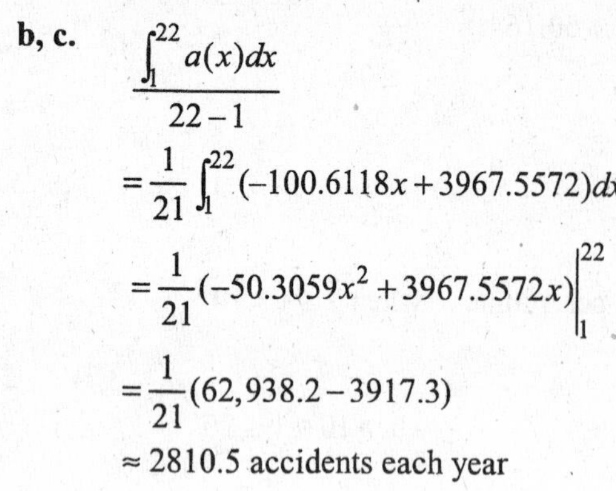

$$\frac{\int_1^{22} a(x)dx}{22-1}$$

$$= \frac{1}{21}\int_1^{22}(-100.6118x + 3967.5572)dx$$

$$= \frac{1}{21}(-50.3059x^2 + 3967.5572x)\Big|_1^{22}$$

$$= \frac{1}{21}(62{,}938.2 - 3917.3)$$

≈ 2810.5 accidents each year

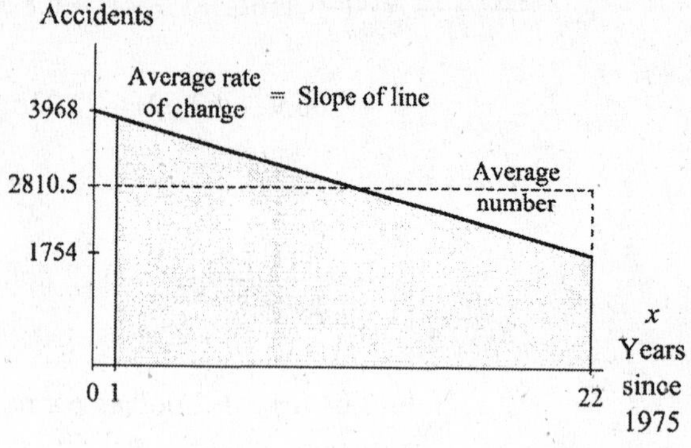

9. a. Average acceleration $= \dfrac{\int_0^{35} a(t)dt}{35-0} = \dfrac{1}{35}\int_0^{35}(0.024t^2 - 1.72t + 22.58)dt$

$$= \frac{1}{35}(0.008t^3 - 0.86t^2 + 22.58t)$$

$$= \frac{1}{35}(79.8 - 0) = 2.28 \text{ ft/sec}^2$$

b. $v(t) = \int_0^t a(x)dx = 0.008t^3 - 0.86t^2 + 22.58t$ feet per second after t seconds

Average velocity $= \dfrac{\int_0^{35} v(t)dt}{35} = \dfrac{1}{35}\int_0^{35}(0.008t^3 - 0.86t^2 + 22.58t)dt$

$$\approx \frac{1}{35}(0.002t^4 - 0.287t^3 + 11.29t^2)\Big|_0^{35}$$

$$\approx \frac{1}{35}(4540.67 - 0) \approx 129.7 \text{ feet/second}$$

c. $s(35) = \int_0^{35} v(t)dt$

This is the same integral that was evaluated in part *b*, so the distance is approximately 4540.7 ft.

d. It would have traveled (129.7 ft/sec)(35 sec) ≈ 4540.7 ft (using unrounded values).

e.

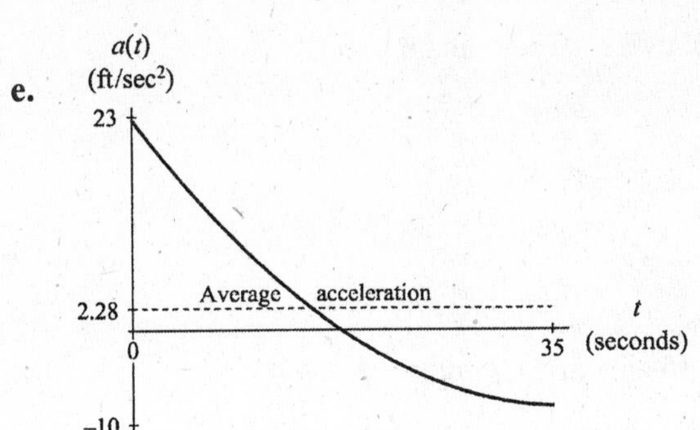

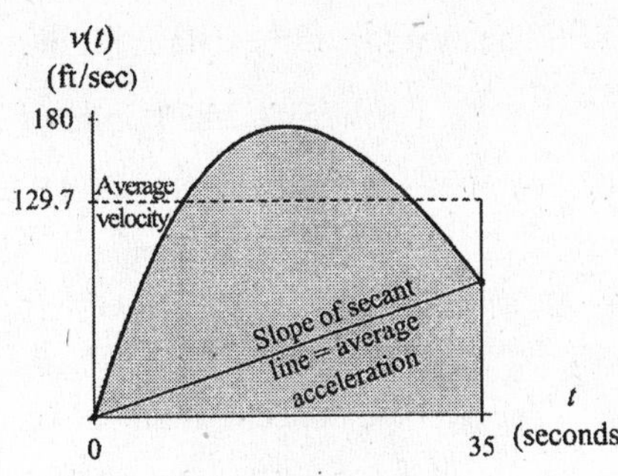

Area of shaded region
= Distance traveled
= Area of rectangle with height 129.7 ft/sec and width 35 seconds

11. a. $V(t) = 1.033t + 138.413$ meters per second t microseconds after the experiment began.

b. $$\text{Average speed} = \frac{\int_{10}^{60} V(t)dt}{60-10} \approx \frac{1}{50}(0.517t^2 + 138.43t)\Big|_{10}^{60}$$

$$\approx \frac{1}{50}(10{,}164.97 - 1435.80) \approx 174.58 \text{ meters per second}$$

13. a. $B(t) = 0.030t^2 - 0.718t + 3.067$ mm Hg per hour t hours after 8 a.m.

b. Note that $B(t)$ is the rate of change, not the actual blood pressure.

$$\text{Average rate} = \frac{\int_0^{12} B(t)dt}{12-0} \approx \frac{1}{12}(0.010t^3 - 0.359t^2 + 3.067t)\Big|_0^{12}$$

$$\approx \frac{1}{12}(2.54 - 0) \approx 0.21 \text{ mm Hg per hour}$$

c. Let $D(t)$ represent the diastolic blood pressure t hours after 8 a.m. Then $D(t) = \int B(t)dt \approx 0.010t^3 - 0.359t^2 + 3.067t + C$. Solving $D(4) = 95$ mm Hg gives $C \approx 87.831$, so $D(t) \approx 0.010t^3 - 0.359t^2 + 3.067t + 87.831$.

$$\text{Average blood pressure} = \frac{\int_0^{12} D(t)}{12}$$

$$\approx \frac{1}{12}\int_0^{12}(0.010t^3 - 0.359t^2 + 3.067t + 87.831)$$

$$\approx \frac{1}{12}\left(0.0025t^4 - 0.120t^3 + 1.534t^2 + 87.831t\right)\Big|_0^{12}$$

$$\approx \frac{1}{12}(1120.3) \approx 93.4 \text{ mm Hg}$$

15. $$\text{Average rate} = \frac{\int_{56}^{100} P(t)dt}{100-56} = \frac{1}{44}\int_{56}^{100}[0.0106t - 1.148]dt \approx -0.32 \text{ seconds per year.}$$

17. a. The average of the data is highest between 10 a.m. and 6 p.m.

b. $C(x)=\begin{cases} 4.75x+2.833 \text{ ppm} & \text{when } 0\le x<4 \\ 0.536x^2-7.871x+45.200 \text{ ppm} & \text{when } 4\le x\le 12 \\ -5.5x+93.667 \text{ ppm} & \text{when } 12<x\le 16 \end{cases}$

x hours after 6 a.m.

The average concentration between 10 a.m. and 6 p.m. is

$\frac{1}{12-4}\int_4^{12} C(x)dx=\frac{1}{8}\int_4^{12}(0.536x^2-7.871x+45.200)dx\approx 19.4$ ppm.

c. Average concentration

$=\frac{1}{16-0}\int_0^{16} C(x)dx$

$=\frac{1}{16}\left[\int_0^4(4.75x+2.833)dx+\int_4^{12}(0.536x^2-7.871x+45.200)dx+\int_{12}^{16}(-5.5x+93.667)dx\right]$

$\approx\frac{1}{16}(49.332+155.157+66.668)$

$=\frac{1}{16}(271.157)\approx 16.94$ ppm

d. Severe pollution warning.

19. *One possible answer:* Consider the two graphs of a function f shown below, where A is the average value of f from a to b and k is an arbitrary constant.

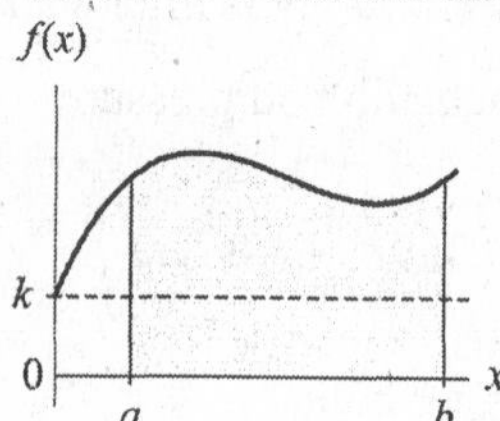

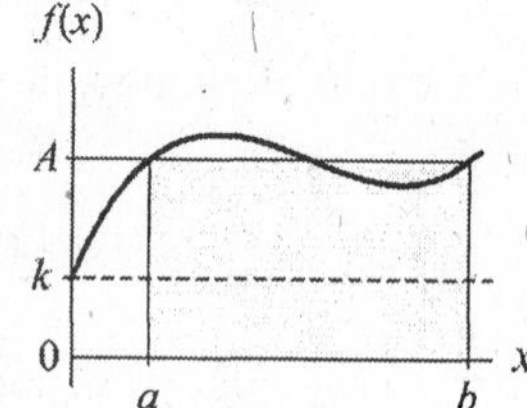

We know that the areas of the two shaded regions are equal. If we remove from each graph the rectangular region with height k and width $b-a$, then the areas of the resulting regions are still equal, because we have removed the same area from each.

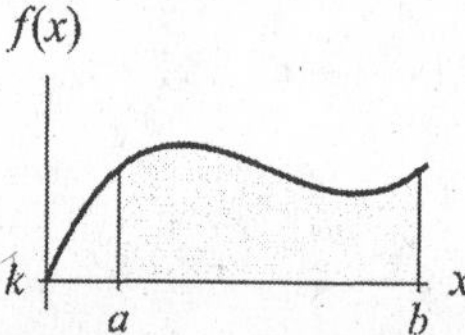

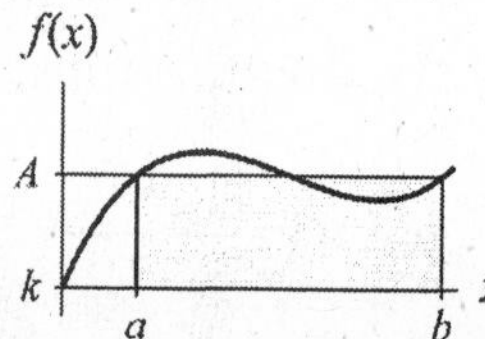

It is true for the graphs shown in this section with vertical axis shown from k rather than from zero that the area of the region between the function and $y=k$ from a to b is the same as the area of the rectangle with height equal to the average value minus k and width equal to $b-a$.

Section 5.6 Integration by Substitution or Algebraic Manipulation

1. Let $u = 2x$ so $du = 2dx$. Then

$$\int 2e^{2x}dx = \int e^u du$$
$$= e^u + C$$
$$= e^{2x} + C$$

3. Not possible using the techniques discussed in this text. **Note: The answer given in the back of the text for this activity is the answer for Activity 2.**

5. Let $u = 1 + e^x$ so $du = e^x dx$. Then

$$\int (1+e^x)^2 e^x dx = \int u^2 du$$
$$= \frac{1}{3}u^3 + C$$
$$= \frac{\left(1+e^x\right)^3}{3} + C$$

7. Let $u = 2^x + 2$ so $du = \ln 2\left(2^x\right)dx$. Then

$$\int \frac{2^x}{2^x+2}dx = \frac{1}{\ln 2}\int \frac{\ln 2\left(2^x\right)}{2^x+2}dx$$
$$= \frac{1}{\ln 2}\int \frac{1}{u}du$$
$$= \frac{1}{\ln 2}\ln u + C$$
$$= \frac{\ln\left(2^x+2\right)}{\ln 2} + C$$

9. Using technology, we estimate the value of the integral as approximately 2.5452.

11. Let $u = \ln x$ so $du = \frac{1}{x}$. Then $\int_2^5 \frac{\ln x}{x}dx = \int_2^5 \ln x \frac{1}{x}dx$ is of the form

$\int u du = \frac{1}{2}u^2 + C$.

Now we back-substitute and evaluate at $x = 2$ and $x = 5$:

$$\left.\frac{(\ln x)^2}{2}\right|_2^5$$
$$= \frac{(\ln 5)^2}{2} - \frac{(\ln 2)^2}{2}$$

(exact)

13. Using technology, we estimate the value of the integral as approximately 3.6609.

15. Let $u = x^2 + 1$ so $du = 2x$. Then $\int_3^4 \frac{2x}{x^2+1}dx = \int_3^4 \frac{1}{x^2+1} 2x dx$ is of the form

$\int \frac{1}{u} du = \ln u + C$.

Now we back-substitute and evaluate at $x = 3$ and $x = 4$.

$\ln(x^2+1)\Big|_3^4 = \ln 17 - \ln 10$ (exact)

17. Using technology, we estimate the value of the integral as approximately 8.7595.

19. $\int_3^4 \frac{x^2+1}{x^2}dx = \int_3^4 \left(1 + \frac{1}{x^2}\right)dx$

$= \left(x - \frac{1}{x}\right)\Big|_3^4$

$= (4 - \frac{1}{4}) - (3 - \frac{1}{3}) = 1\frac{1}{12}$

(exact)

Chapter 5 Concept Review

1. a.

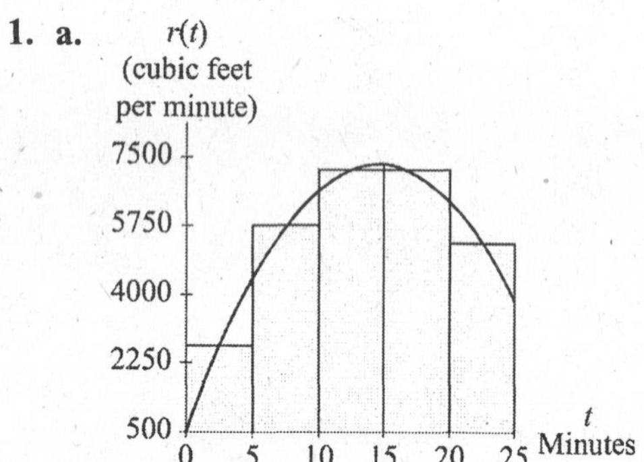

b. The width of each rectangle is $\frac{25-0}{5} = 5$.

Interval	Midpoint t minutes	Rectangle height $r(t)$ cubic feet per minute	Rectangle area (5 minutes)[$r(t)$ ft^3 per minute]→ cubic feet
0 to 5	2.5	2639.5	13,197.5
5 to 10	7.5	5704.5	28,522.5
10 to 15	12.5	7169.5	35,847.5
15 to 20	17.5	7034.5	35,172.5
20 to 25	22.5	5299.5	26,497.5
		Midpoint rectangle area = 139,237.5 cubic feet	

c. In the first 25 minutes that oil was flowing into the tank, approximately 139,238 cubic feet of oil flowed in.

2. a. A quadratic model for the data is
$S(t) = -1.643t^2 + 16.157t + 0.2$ miles per hour t hours after midnight

b.

Number of rectangles	Approximation of area
5	127.131
10	126.869
20	126.803
40	126.786
80	126.782
160	126.781
	Trend ≈ 126.8

The hurricane traveled approximately 126.8 miles.

3. a. The area beneath the horizontal axis represents the amount of weight that the person lost during the diet.

b. The area above the axis represents the amount of weight that the person regained between weeks 20 and 30.

c. Because $-26.7 + 15.4 = -11.3$, the person's weight was 11.3 pounds less at 30 weeks than it was at 0 weeks.

d.

Weight (pounds)
0 10 20 30
Weeks
-11.3
-26.7

e. The graph in part *d* represents the change in the person's weight as a function of the number of weeks since the person began the diet.

4. a. Find the general antiderivative: $R(t) = \int r(t)dt = 10\left(\frac{-3.2}{3}t^3 + \frac{93.3}{2}t^2 + 50.7t\right) + C$

$$= \frac{-32}{3}t^3 + 466.5t^2 + 507t + C$$

Solve for C:

$$R(0) = 5000$$

$$\frac{-32}{3}(0)^3 + 466.5(0)^2 + 507(0) + C = 5000$$

$$C = 5000$$

The specific antiderivative is $R(t) = \frac{-32}{3}t^3 + 466.5t^2 + 507t + 5000$ cubic feet after t minutes

b. $R(10) - R(0) \approx 46{,}053.3 - 5000 = 41{,}053.3$ ft^3

c. Use technology to solve $R(t) = 150{,}000$, which gives $t \approx 27.55$ or $t \approx 31.73$. The tank will be full after approximately 27.6 minutes.

5. a. $\int_0^{2.75} a(x)dx = \int_0^{2.75} 840(1.08763^x)dx$

$$= \left.\frac{840(1.08763^x)}{\ln 1.0763}\right|_0^{2.75}$$

$$\approx 12{,}598.475 - 9999.879 \approx \$2598.60$$

b. At the end of the third quarter of the third year, the \$10,000 had increased by \$2598.60 so that the total value of the investment was \$12,598.60.

6. $\int_{79}^{88} [m(t) - w(t)]\, dt \approx \$123$ thousand

Between the beginning of 1980 and the end of 1988, a man earning the average full-time wage would have earned approximately \$123,000 more than a woman earning the average full-time wage.

Chapter 6
Analyzing Accumulated Change: Integrals in Action

Section 6.1 Perpetual Accumulation and Improper Integrals

1. $$\int_0^\infty 3e^{-0.2t}dt = \lim_{N\to\infty}\int_0^N 3e^{-0.2t}dt$$
$$= \lim_{N\to\infty}\left(\frac{3}{-0.2}e^{-0.2t}\right)\Big|_0^N$$
$$= \lim_{N\to\infty}\left(\frac{3}{-0.2}e^{-0.2N} - \frac{3}{-0.2}e^{-0.2(0)}\right)$$
$$= \lim_{N\to\infty}\left(\frac{3}{-0.2}e^{-0.2N} + 15\right) = 0 + 15 = 15$$

3. $$\int_{10}^\infty 3x^{-2}dx = \lim_{N\to\infty}\int_{10}^N 3x^{-2}dx$$
$$= \lim_{N\to\infty}\left(-3x^{-1}\Big|_{10}^N\right)$$
$$= \lim_{N\to\infty}\left(\frac{-3}{N} - \frac{-3}{10}\right)$$
$$= 0 + \frac{3}{10} = \frac{3}{10} = 0.3$$

5. $$\int_{-\infty}^{-10} 4x^{-3}dx = \lim_{N\to-\infty}\int_N^{-10} 4x^{-3}dx$$
$$= \lim_{N\to-\infty}\left(-2x^{-2}\Big|_N^{-10}\right)$$
$$= \lim_{N\to-\infty}\frac{-2}{(-10)^2} - \frac{-2}{N^2}$$
$$= -0.2 - 0 = -0.2$$

7. $$\int_{0.36}^\infty 9.6x^{-0.432}dx = \lim_{N\to\infty}\int_{0.36}^N 9.6x^{-0.432}dx$$
$$= \lim_{N\to\infty}\left(\frac{9.6}{0.568}x^{0.568}\Big|_{0.36}^N\right)$$
$$= \lim_{N\to\infty}\left(\frac{9.6}{0.568}N^{0.568} - \frac{9.6}{0.568}\left(0.36^{0.568}\right)\right) \to \infty$$

This integral diverges.

9. $$\int_2^{\infty} \frac{2x}{x^2+1}dx = \lim_{N\to\infty} \int_2^{N} \frac{2x}{x^2+1}dx$$
$$= \lim_{N\to\infty}\left[\ln(x^2+1)\Big|_2^N\right]$$
$$= \lim_{N\to\infty}\left[\ln(N^2+1)-\ln 5\right] \to \infty$$

This integral diverges.

11. $$\int_a^{\infty}[f(x)+k]dx = \int_a^{\infty} f(x)dx + \int_a^{\infty} k dx$$

The second integral can be thought of as the area of a rectangle with height k and infinite width. The second integral diverges, so the original integral diverges.

13. a. Amount after 100 years = $\int_0^{100} -1.55(0.9999999845^t)\cdot 10^{-6}dt$

$$= \frac{-1.55(0.9999999845^t)\cdot 10^{-6}}{\ln 0.9999999845}\Bigg|_0^{100}$$
$$= \frac{-1.55(0.9999999845^{100})\cdot 10^{-6}}{\ln 0.9999999845} - \frac{-1.55(0.9999999845^{0})\cdot 10^{-6}}{\ln 0.9999999845}$$
$$\approx -0.0002 \text{ milligram}$$

After 100 years, only 0.0002 mg of ^{238}U will have decayed.

Amount after 1000 years

$$= \int_0^{1000} -1.55(0.9999999845^t)\cdot 10^{-6}dt = \frac{-1.55(0.9999999845^t)\cdot 10^{-6}}{\ln 0.9999999845}\Bigg|_0^{1000}$$
$$= \frac{-1.55(0.9999999845^{1000})\cdot 10^{-6}}{\ln 0.9999999845} - \frac{-1.55(0.9999999845^{0})\cdot 10^{-6}}{\ln 0.9999999845}$$
$$\approx -0.0015 \text{ milligram}$$

After 1000 years, only 0.0015 mg of ^{238}U will have decayed.

b. $$\int_0^{\infty} -1.55(0.9999999845^t)\cdot 10^{-6}dt = \lim_{N\to\infty}\int_0^{N} -1.55(0.9999999845^t)\cdot 10^{-6}dt$$
$$= \lim_{N\to\infty}\left(\frac{-1.55(0.9999999845^t)\cdot 10^{-6}}{\ln 0.9999999845}\Bigg|_0^{N}\right)$$
$$= \lim_{N\to\infty}\left[\frac{-1.55(0.9999999845^N)\cdot 10^{-6}}{\ln 0.9999999845} - \frac{-1.55(0.9999999845^{0})\cdot 10^{-6}}{\ln 0.9999999845}\right]$$
$$= 0 - \frac{-1.55(0.9999999845^{0})\cdot 10^{-6}}{\ln 0.9999999845} = -99.99999923 \approx -100 \text{ milligrams}$$

Eventually, all of the ^{238}U will decay.

15. a. Solving $150 = 499.589(0.958^p)$ yields $p_0 \approx \$28.04$.

b. (Note that we use the unrounded value of p_0 in calculations)

$$C = q_0 p_0 + \int_{p_0}^{\infty} D(p)dp \approx 150(28.04) + \int_{28.04}^{\infty} 499.589(0.958^p)dp$$

$$= 4206 + \lim_{N\to\infty} \int_{28.04}^{N} 499.589(0.958^p)dp = 4206 + \lim_{N\to\infty}\left[\left.\frac{499.589(0.958^p)}{\ln 0.958}\right|_{28.04}^{N}\right]$$

$$= 4206 + \lim_{N\to\infty}\left[\frac{499.589(0.958^N)}{\ln 0.958} - \frac{499.589(0.958^{28.04})}{\ln 0.958}\right]$$

$$= 4206 + 0 - \frac{499.589(0.958^{28.04})}{\ln 0.958} \approx \$7702 \text{ thousand}$$

Consumers are willing and able to spend \$7.7 million for 150,000 books.

17. $\int_0^{\infty} 0.1e^{-0.1x}dx = \lim_{N\to\infty}\int_0^{N} 0.1e^{-0.1x}dx = \lim_{N\to\infty}\left(\left.-e^{-0.1x}\right|_0^N\right)$

$$= \lim_{N\to\infty}\left[-e^{-0.1N} - (-e^0)\right]$$

$$= \lim_{N\to\infty} -e^{-0.1N} + \lim_{N\to\infty}(e^0) = 0 + 1 = 1$$

Section 6.2 Streams in Business and Biology

1. a. i. $R(m) = 0.2\left(\frac{\$47{,}000}{12}\right) = \$783.33$ per month

ii. $R(m) = 0.2\left(\frac{47{,}000}{12} + 100m\right)$

$= 783.33 + 20m$ dollars/month after m months

iii. $R(m) = 0.2\left(\frac{47{,}000}{12}(1.005^m)\right)$

$= 7833.33(1.005^m)$

dollars per month after m months

b. i. $\sum_{m=0}^{59} \frac{9400}{12}\left(1+\frac{0.05}{12}\right)^{60-m} = \$53{,}493.40$

ii. $\sum_{m=0}^{59}\left(\frac{9400}{12} + 20m\right)\left(1+\frac{0.05}{12}\right)^{60-m}$

$= \$92{,}082.72$

iii. $\sum_{m=0}^{59} \frac{9400}{12}(1.005^m)\left(1+\frac{0.05}{12}\right)^{60-m}$

$= \$61,818.49$

The first option is the only one that will not result in the amount needed for the down payment.

3. $\int_0^4 (0.125)(17.628)(1.05^t)e^{0.07(4-t)}\,dt = \$11.2$ billion

5. a. $\int_0^7 (177.26)(0.03)e^{0.088(7-x)}\,dx \approx \$51.5$ billion

b. $P = \frac{51.5}{e^{0.088(7)}} \approx \$27.8$ billion

7. a. Because \$500 per month is equivalent to \$6000 per year, use $R(t) = 6000$ dollars per year.

Future value $= \int_0^T R(t)e^{r(T-t)}\,dt = \int_0^6 6000e^{0.0634(6-t)}\,dt = \int_0^6 6000e^{0.0634(6)}e^{-0.0634t}\,dt$

$$= \left.\frac{6000e^{0.0634(6)}e^{-0.0634t}}{-0.0634}\right|_0^6 \approx 43,804.70$$

The investments will be worth approximately \$43,804.70.

b. $\sum_{m=0}^{71} 500\left(1+\frac{0.0634}{12}\right)^{72-m} \approx 43,896.84$

The investments will be worth approximately \$43,896.84.

c. It makes more sense to depict the money entering the account discretely (b) instead of continuously (a). This is generally true when discussing individuals instead of corporations.

9. a. $r(q) = 82.1(1.05^q)(0.15)$ billion dollars per quarter q quarters after the third quarter of 2002.

b. $R(q) = 82.1(1.05^q)(0.15)(1.09^{16-q})$ million dollars per quarter for money invested q quarters after the third quarter of 2002.

c. If the investment begins with the 4th quarter 2002 profits, then the initial investment is based on a profit of $(82.1)(1.05) = \$86.205$ billion. Thus we çalculate

$\sum_{q=0}^{15} 86.205(1.05^q)(0.15)(1.09^{16-q}) \approx 629.8$ billion dollars

(Note that this value does not include the contribution made at the end of year 2006.)

11. a. The income is earned at a rate of

$$R(t) = \left(\frac{36{,}400 \text{ liters}}{3 \text{ years}}\right)\left(\frac{\$0.80}{1 \text{ liter}}\right)$$

$$= \$9706.67 \text{ per year}$$

Future value

$$= \int_0^T R(t)e^{r(T-t)}dt$$

$$= \int_0^7 (9706.67)e^{0.045(7-t)}dt$$

$$= \int_0^7 (9706.67)e^{0.045(7)}e^{-0.045t}dt$$

$$= \left.\frac{(9706.67)e^{0.045(7)}e^{-0.045t}}{-0.045}\right|_0^7$$

$$= \left.\frac{(9706.67)e^{0.045(7-t)}}{-0.045}\right|_0^7$$

$$\approx -215{,}703.70 - (-295{,}570.01)$$

$$= 79{,}866.31 \text{ dollars}$$

Pepsi will make approximately \$79,866.

b. Let P = present value. Since $Pe^{rt} = Pe^{0.045(7)} \approx 79{,}866.31$, the present value is $P \approx \frac{79{,}866.31}{e^{0.045(7)}} \approx 58{,}285$ dollars. Pepsi would have had to invest approximately \$58,285.

13. a. Let P = present value, and use the result of Activity 7a.

$$Pe^{rt} = \text{future value}$$

$$\text{P}e^{0.0634(6)} \approx 43{,}804.70$$

$$\text{P} \approx \frac{43{,}804.70}{e^{0.0634(6)}} \approx 29{,}944.36$$

You would need to invest approximately \$ 29,944 .

Note: This amount can also be determined by calculating $\int_0^6 6000e^{-0.0634t}dt$.

b. Let P = present value, and use the result of Activity 7b.

$$P\left(1+\frac{0.0634}{12}\right)^{12(6)} \approx 43{,}804.70$$

You would need to invest approximately \$30,037.41.

c. It makes more sense to depict the money entering the account discretely (b) instead of continuously (a). This is generally true when discussing individuals instead of corporations.

15. a. Present value $= \int_0^T R(t)e^{-rt}\,dt$

$$= \int_0^{20} 850e^{-0.15t}\,dt = \left.\frac{850e^{-0.15t}}{-0.15}\right|_0^{20}$$

$\approx -282.2 - (-5666.7)$

$= \$5384.5$ million

The 20-year present value is approximately \$5384.5 million, or \$5.4 billion.

b. Present value $= \int_0^T R(t)e^{-rt}\,dt$

$$= \int_0^{20} 850e^{-0.13t}\,dt = \left.\frac{850e^{-0.13t}}{-0.13}\right|_0^{20}$$

$\approx -485.63 - (-6538.46)$

$\approx \$6052.8$ million

The 20-year present value is approximately \$6052.8 million, or \$6.1 billion.

c. Use $R(t) = 850(1.1^t)$ million dollars per year.

Present value $= \int_0^T R(t)e^{-rt}\,dt$

$$= \int_0^{20} 850(1.1^t)e^{-0.14t}\,dt$$

$$= \int_0^{20} 8.5(1.1e^{-0.14})^t\,dt$$

$$= \left.\frac{8.5(1.1e^{-0.14})^t}{\ln(1.1e^{-0.14})}\right|_0^{20}$$

$\approx -7781.1 - (-19{,}020.0)$

$= \$11{,}238.9$ million

The 20-year present value is approximately \$11,238.9 million, or \$11.2 billion.

Note: We have assumed that the annual returns given refer to APRs, compounded continuously. If they are interpreted as APYs, the answers to parts *a*, *b*, and *c* are \$5410.2 million, \$6351.3 million, and \$12,148.5 million, respectively.

17. a. Use $R(t) = 2(0.95^t)$ billion dollars per year.

Present value $= \int_0^T R(t)e^{-rt}\,dt$

$$= \int_0^{10} 2(0.95^t)e^{-0.2t}\,dt = \int_0^{10} 2(0.95e^{-0.2})^t\,dt = \left.\frac{2(0.95e^{-0.2})^t}{\ln(0.95e^{-0.2})}\right|_0^{10}$$

$\approx -0.645 - (-7.959) = \$7.314$ billion

The 10-year present value is approximately \$7.3 billion. This is less than the \$8.1 billion offered by CSX.

b. Use $R(t) = 1.2(1.02^t)$ billion dollars per year.

$$\text{Present value} = \int_0^T R(t)e^{-rt}\,dt$$
$$= \int_0^{10} 1.2(1.02^t)e^{-0.2t}\,dt$$
$$= \int_0^{10} 1.2(1.02e^{-0.2})^t\,dt$$
$$= \left.\frac{1.2(1.02e^{-0.2})^t}{\ln(1.2e^{-0.2})}\right|_0^{10}$$
$$\approx -1.0986 - (-6.6594)$$
$$\approx \$5.561 \text{ billion}$$

The 10-year present value is approximately \$5.6 billion.

c. Different companies will have different ideas as to how another company of themselves will perform.

Note: We have assumed that the annual returns given refer to APRs, compounded continuously. If they are interpreted as APYs, the answers to part *a* and *b* are \$7.7 billion (less than CSX offered), and \$5.930 billion, respectively.

19. Use $R(t) = 1.2(1.06^t)$ million dollars per year.

$$\text{Present value} = \int_0^T R(t)e^{-rt}\,dt$$
$$= \int_0^5 1.2(1.06^t)e^{-0.12t}\,dt$$
$$= \int_0^5 1.2(1.06e^{-0.12})^t\,dt$$
$$= \left.\frac{1.2(1.06e^{-0.12})^t}{\ln(1.06e^{-0.12})}\right|_0^5$$
$$\approx -14.28 - (-19.44)$$
$$\approx \$5.16 \text{ million}$$

The capital value, or 5-year present value, is approximately \$5.2 million.

21. Answers will vary, depending on the current year. The given solution is based on the end of 2001.

a. $(12 \text{ million})(0.83^{22}) \approx 0.20$ million terns

b. Use $r(t) = 2.04$ million terns per year, and $s = 0.83$. The desired expression is $r(t)s^{22-t} = 2.04(0.83)^{22-t}$ million terns hatched t years after 1979.

c. Future value $= Ps^b + \int_0^b r(t)s^{b-t}\,dt$

$$= 12(0.83^{22}) + \int_0^{22} 2.04(0.83^{22-t})\,dt$$

$$\approx 0.20 + \int_0^{22} 2.04(0.83^{22})(0.83^{-1})^t\,dt$$

$$= \frac{2.04(0.83^{22})(0.83^{-1})^t}{\ln(0.83^{-1})}\Bigg|_0^{22} + 0.20 = \frac{2.04(0.83^{22-t})}{-\ln 0.83}\Bigg|_0^{22} + 0.20$$

$$\approx 10.95 - 0.18 + 0.20 = 10.97 \text{ million terns}$$

There are approximately 10.97 million sooty terns at the end of 2001.

23. a. $(200 \text{ thousand})(0.67^{50}) \approx 4\cdot 10^{-7}$ thousand ≈ 0

None of the current population will still be alive.

b. Use $r(t) = 60 - 0.5t$ thousand seals per year, and $s = 0.67$. The desired expression is $r(t)s^{50-t} = (60-0.5t)(0.67^{50-t})$ thousand seals born t years from now.

c. Use technology to evaluate the definite integral.

Future value $= Ps^b + \int_0^b r(t)s^{b-t}\,dt \approx 0 + \int_0^{50} (60-0.5t)(0.67)^{50-t}\,dt \approx$ 90.5 thousand seals

There will be approximately 90.5 thousand seals.

24. *One possible answer:* Discrete income streams are depicted by set times when money enters into an account. This is usually true of individuals making deposits. Continuous income streams are represented by money flowing into an account at a steady rate. There are no set times when money enters the account. It is always flowing into the account. This is usually assumed to be true of large businesses and corporations.

Section 6.3 Integrals in Economics

1. a. The demand function **b.** The supply function

c. The producers' surplus **d.** The consumers' surplus

3. a. To find the price P above which consumers will purchase none of the goods or services, either find the smallest positive value for which the demand function is zero, $D(p) = 0$, or, if $D(p)$ is never exactly zero but approaches zero as p increases without bound, then let $P \to \infty$.

b. The supply function S is a piecewise continuous function with the first piece being the 0 function. The value p at which $S(p)$ is no longer 0 is the shutdown price. The shutdown point is $(p_1, S(p_1))$.

c. The market equilibrium price p_0 can be found as the solution to $S(p) = D(p)$. That is, it is the price at which demand is equal to supply. The equilibrium point is the point $(p_0, D(p_0)) = (p_0, S(p_0))$.

5. a. $q_0 \approx 27.5$ thousand units

b, c.

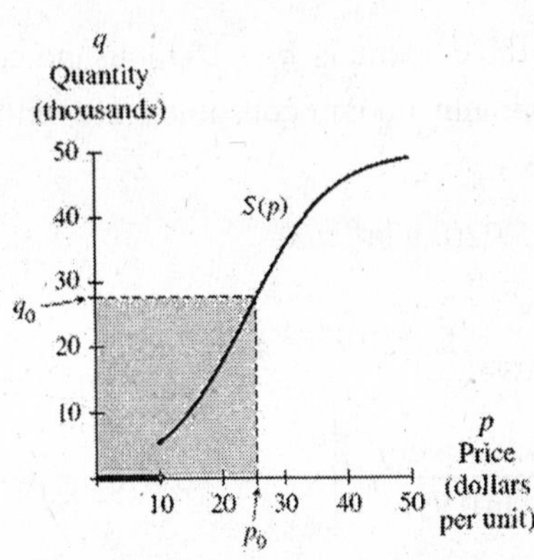

(a) Mathematician's viewpoint

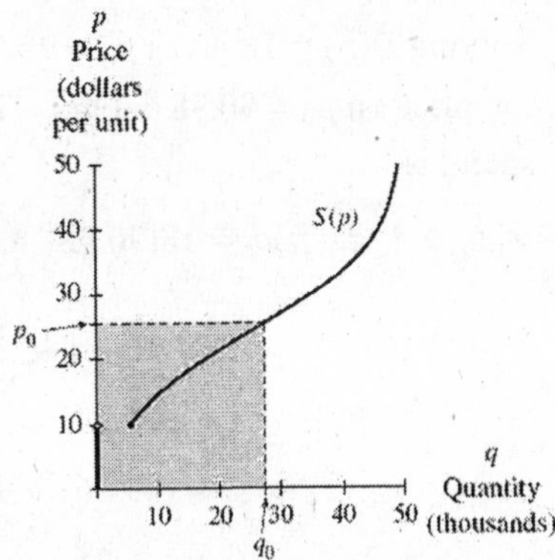

(b) Economist's viewpoint

7. a.

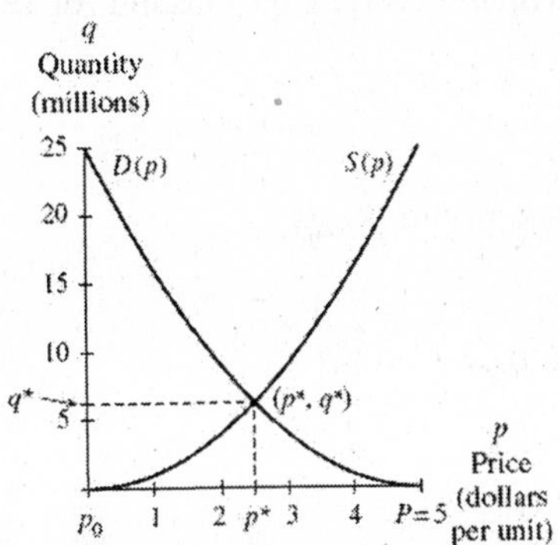

Mathematician's viewpoint

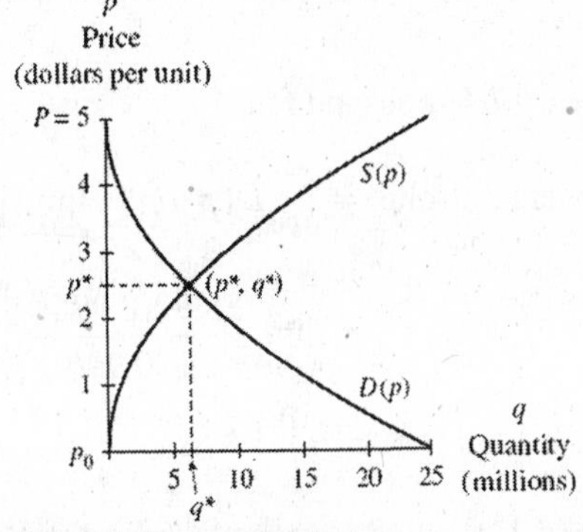

Economist's viewpoint

$p^* \approx 2.5$ dollars per unit, $q^* \approx 2.5$ million units, $p_1 \approx 0$ dollars per unit, $P \approx 0$ dollars per unit

b, c.

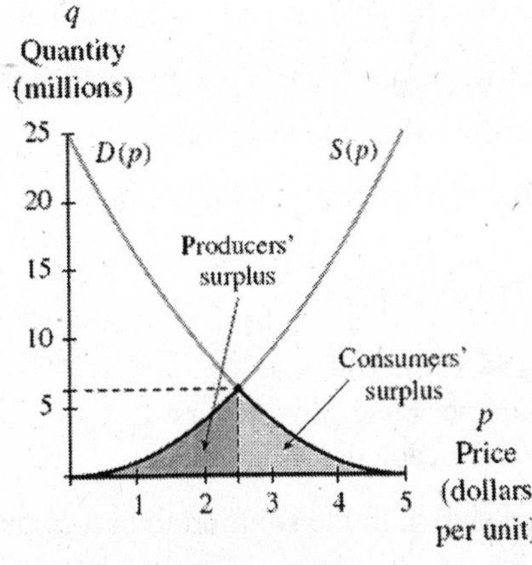

Mathematician's viewpoint

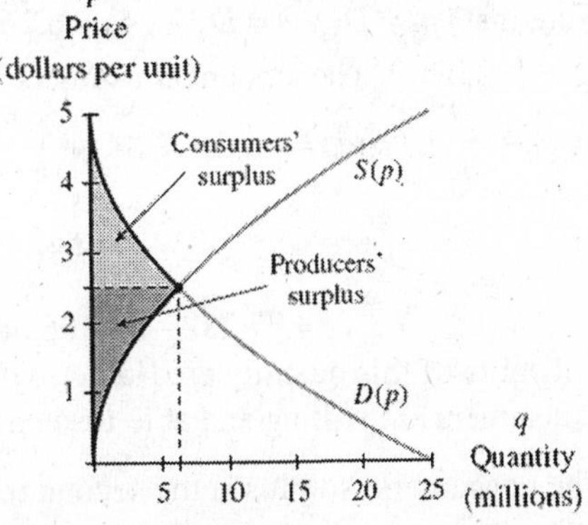

Economist's viewpoint

d. Total Social Gain $= \int_0^{2.5} S(p)dp + \int_{2.5}^{5} D(p)dp$

9. a. D is an exponential demand function and so does not have a finite value p at which $D(p) = 0$. Thus the model does not indicate a price above which consumers will purchase none of the goods or services.

b. Solving $D(p) = 18$ gives $p_0 \approx 90.98$, so the demand is $q_0 = 18$ thousand ceiling fans when the price is $p_0 \approx 90.98$ dollars. The maximum amount consumers are willing and able to spend is

$$p_0 q_0 + \int_{p_0}^{P} D(p)dp \approx 18(90.98) + \int_{90.98}^{\infty} 25.92(0.996^p)dp$$

$$= 1637.61 + \lim_{N\to\infty} \int_{90.98}^{N} 25.92(0.996^p)dp$$

$$= 1637.61 + \lim_{N\to\infty} \left.\frac{25.92(0.996^p)}{\ln 0.996}\right|_{90.98}^{N} \approx 1637.61 + 0 - (-4490.99)$$

$$= 6128.60$$

The units of this quantity are (thousands of fans)(dollars per fan) = thousands of dollars. Consumers are willing and able to spend approximately \$6128.6 thousand for 18 thousand fans.

c. $D(100) \approx 17.4$ thousand fans

d. Consumers' surplus $= \int_{100}^{\infty} D(p)dp = \lim_{N\to\infty} \int_{100}^{N} 25.92(0.996^p)\,dp$

$$= \lim_{N\to\infty} \left.\frac{25.92(0.996^p)}{\ln 0.996}\right|_{100}^{N} \approx 0 - (-4331.5)$$

$$= 4331.5$$

As in part *b*, the units are thousands of dollars. The consumers' surplus is approximately \$4331.5 thousand.

11. a. $D(p) = 0.025p^2 - 1.421p + 19.983$ lanterns when the market price is \$$p$ per lantern

b. Note that $q_0 = D(p_0) = D(12.34) \approx 6.263$. Also, $D(p) = 0$ when $p \approx 25.701$ or $p \approx 31.075$, so $P \approx 25.701$. The amount consumers are willing and able to spend is

$$p_0 q_0 + \int_{p_0}^{P} D(p)dp \approx 6.263(12.34) + \int_{12.34}^{25.701} (0.025p^2 - 1.421p + 19.983)dp$$

$$\approx 77.288 + \left.(0.008p^3 - 0.710p^2 + 19.983p)\right|_{12.34}^{25.701}$$

$$\approx 77.288 - 0 + 186.000 - 154.106 = 109.18$$

The units of this quantity are (lanterns)(dollars per lantern) = dollars. Consumers are willing and able to spend \$109.18 each day.

c. The consumers' surplus is the second integral shown in the solution to part *b*; that is,

$\int_{p_0}^{P} D(p)dp \approx 186.000 - 154.106 = 31.894$ dollars or \$31.89.

13.a. First, we find an expression for elasticity:

$$\eta = \left|\frac{p \cdot D'(p)}{D(p)}\right|$$

$$= \left|\frac{p \cdot 25.92(0.996^p)\ln(0.996)}{25.92(0.996^p)}\right|$$

Solving $\eta = 1$, we find that $p = 249.50$. Unit elasticity occurs when chairs are priced at $249.50 per fan.

b. We calculate η for values of p on either side of 249.5. $p = 225$ yields 0.902 and $p = 275$ yields 1.102. Therefore, for prices less than $249.50 per chair demand is inelastic and for prices above $249.50 per chair, demand is elastic.

15.a. First, we find an expression for elasticity:

$$\eta = \left|\frac{p \cdot D'(p)}{D(p)}\right|$$

$$= \left|\frac{p \cdot (0.05p - 1.421)}{0.025p^2 - 1.421p + 19.983}\right|$$

Solving $\eta = 1$, we find that $p = 9.33$. Unit elasticity occurs when lanterns are priced at $9.33 per lantern.

b. We calculate η for values of p on either side of 9.33. $p = 9$ yields 0.948 and $p = 10$ yields 1.113. Therefore, for prices less than $9.33 per lantern demand is inelastic and for prices above $9.33 per lantern, demand is elastic.

17. a. $S(40) = 0.024(40)^2 - 2(40) + 60 = 18.4$ thousand answering machines

$S(150) = 0.024(150)^2 - 2(150) + 60 = 300$ thousand answering machines

Producers will supply 18,400 answering machines at $40, and 300,000 answering machines at $150.

b. Because $S(99.95) = 99.86006$ thousand answering machines, the producers' revenue is ($99.95 per answering machine)(99.86006 thousand answering machines) ≈ $9981 thousand, or $9,981,000.

$$\text{Producers' surplus} = \int_{p_1}^{p_0} S(p)dp$$

$$= \int_{20}^{99.95} (0.024p^2 - 2p + 60)dp$$

$$= 0.008p^3 - p^2 + 60p\Big|_{20}^{99.95} \approx 3995.0 - 864.0 = 3131$$

The units of this quantity are (dollars per answering machine)(thousands of answering machines) = thousands of dollars, so the producers' surplus is approximately $3131 thousand, or $3,131,000.

19. a. $S(p) = \begin{cases} 0 \text{ hundred prints} & \text{when } p < 5 \\ 0.300p^2 - 3.126p + 10.143 \text{ hundred prints} & \text{when } p \geq 5 \end{cases}$

where p hundred dollars is the price of a print.

b. Solving $S(p) = 5$ gives $p \approx 8.3712$ hundred dollars. Producers will supply 500 prints at a price of \$837.12.

c. Because $S(6.3) \approx 2.358$ hundred prints, the producers' revenue is (\$630 per print)(235.8 prints) $\approx$ \$148,500.

$$\text{Producers' surplus} = \int_{p_1}^{p_0} S(p)dp \approx \int_5^{6.3} (0.300p^2 - 3.126p + 10.143)dp$$

$$= 0.100p^3 - 1.563p^2 + 10.143p\Big|_5^{6.3} \approx 26.875 - 24.123 = 2.732$$

The units of this quantity are
(hundreds of dollars per print)(hundreds of prints) = tens of thousands of dollars, so the producers' surplus is approximately \$27,300.

21. a. Because the solutions to $D(p) = 0$ are $p \approx -41.20$ and $p \approx 20.577$, we use $P \approx 20.577$. Solving $D(p) = 20$ gives $p \approx 20.252$, so we use $p_0 \approx 20.252$ and $q_0 = 20$.
The amount consumers are willing and able to spend is

$$p_0 q_0 + \int_{p_0}^{P} D(p)dp \approx 20(20.252) + \int_{20.252}^{20.577} (-1.003p^2 - 20.689p + 850.375)dp$$

$$= 405.04 + \left(\frac{-1.003p^3}{3} - 10.3445p^2 + 850.375p\right)\Bigg|_{20.252}^{20.577}$$

$$\approx 405.04 + 10{,}205.27 - 10{,}202.02$$

$$= 408.30$$

The units of this quantity are
(hundreds of dollars per sculpture)(sculptures) = hundreds of dollars, so consumers are willing and able to spend approximately \$408.3 hundred or \$40,830.

b. Because $D(5) = 850.375$ and $S(5) = 297.157$, the quantity supplied at \$500 per sculpture is 297 sculptures, and supply will not exceed demand.

c. Solving $D(p) = S(p)$ gives $p \approx 13.21$ hundred dollars, so the equilibrium price is approximately \$1321 per sculpture. The positive solution to $D(p) = 0$ is $p \approx 20.58$ hundred dollars, so consumers will not purchase when the price is over \$2058 and we use $P \approx 20.58$.
The total social gain is

$$\int_{p_1}^{p^*} S(p)dp + \int_{p^*}^{P} D(p)dp$$

$$\approx \int_{4.5}^{13.21} (0.256p^2 + 8.132p + 250.097)dp + \int_{13.21}^{20.58} (-1.003p^2 - 20.689p + 850.375)dp$$

$$= \left(\frac{0.256p^3}{3} + 4.066p^2 + 250.097p\right)\Bigg|_{4.5}^{13.21} + \left(\frac{-1.003p^3}{3} - 10.3445p^2 + 850.375p\right)\Bigg|_{13.21}^{20.58}$$

$\approx (4209.07 - 1215.55) + (10{,}205.27 - 8656.63)$
≈ 4542.15 (using unrounded values)

The units of this quantity are hundreds of dollars, so the total social gain is approximately \$4542.15 hundred or \$454,215.

23. a. $D(p) = 499.589(0.958086^p)$ thousand books when the market price is \$$p$ per book.

b. $S(p) = \begin{cases} 0 \text{ thousand books} & \text{when } p < 18.97 \\ 0.532p^2 - 20.060p + 309.025 \text{ thousand books} & \text{when } p \ge 18.97 \end{cases}$

where \$$p$ is the price of a book.

c. Solving $D(p) = S(p)$ gives an equilibrium price of $p^* \approx \$27.15$ per book. Since $D(p^*) = S(p^*) \approx 156.2$ thousand books, approximately 156.2 thousand books will be supplied and demanded.

d. Total social gain $= \int_{p_1}^{p^*} S(p)dp + \int_{p^*}^{P} D(p)dp$

$$\approx \int_{18.97}^{27.15} (0.532p^2 - 20.060p + 309.025)d + \int_{27.15}^{\infty} 499.589(0.958086^p)dp$$

$$\approx (0.177p^3 - 10.030p^2 + 309.025p)\Big|_{18.97}^{27.15} + \lim_{N\to\infty}\left(\frac{499.589(0.958086^p)}{\ln 0.958086}\right)\Bigg|_{27.15}^{N}$$

$\approx 4542.71 - 3462.26 + [0 - (-3648.16)]$
≈ 4728.62 (using unrounded values)

The units of this quantity are (dollars per book)(thousands of books) = thousands of dollars, so the total social gain is approximately \$4728.6 thousand.

Section 6.4 Probability Distributions and Density Functions

1. a. There is a 46% chance that any telephone call made on a computer software technical support line will be 5 minutes or more.

b. The likelihood that any two cars on a certain two-lane road are less than 7 feet apart is approximately 25%.

c. New Orleans will receive between 2 and 4 inches of rain during the month of March 15% of the time.

3. a. $f(x) \ge 0$ for all x.

$$\int_{-\infty}^{\infty} f(x)dx = \int_0^1 1.5\left(1 - x^2\right)dx = 1.5\left(x - \tfrac{1}{3}x^3\right)\Big|_0^1 = 1.5\left(1 - \tfrac{1}{3}\right) = 1$$

Yes, f is a probability density function.

b. $h(x) \ge 0$ for all x.

$$\int_{-\infty}^{\infty} h(x)dx = \int_0^1 6\left(x - x^2\right)dx = 6\left(\tfrac{1}{2}x^2 - \tfrac{1}{3}x^3\right)\Big|_0^1 = 6\left(\tfrac{1}{2} - \tfrac{1}{3}\right) = 1$$

Yes, h is a probability density function.

c. $r(t) \geq 0$ for all t.

The area between the graph of r and the t-axis is $\frac{1}{2}(0.5)(1.2)+\frac{1}{2}(1)(1.2)=0.3+0.6 \neq 1$. No, r is not a probability density function.

d. Because $s(c)$ is negative for some values of c, s is not a probability density function.

5. a. $P(x<1) = \int_0^1 y(x)dx = \int_0^1 0.32x dx = (0.16x^2)\Big|_0^1 = 0.16 - 0 = 0.16$

b. $\mu = \int_{-\infty}^{\infty} xy(x)dx = \int_0^{2.5} 0.32x^2 dx = \left(\frac{0.32}{3}x^3\right)\Big|_0^{2.5} = \frac{0.32}{3}(2.5^3 - 0) \approx 1.67$

The mean is approximately 167 gallons.

c.

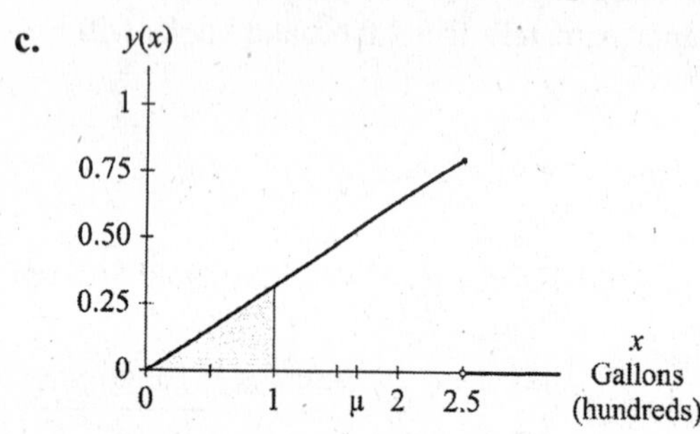

7. Because the values of $f(x)$ are all non-negative and $\int_{-\infty}^{\infty} f(x)dx = 1$, the probability (which is the area between the graph of f and the input axis) must always be between 0 and 1. Another explanation is that the probability of some occurrence is the proportion of times it is expected to happen, and all proportions are fractions between 0 and 1.

9. a. $P(20<t<30) = \int_{20}^{30} 0.2e^{-0.2t}dt = -e^{-0.2t}\Big|_{20}^{30} = -e^{-6} + e^{-4} \approx 0.016$

The probability that successive arrivals are between 20 and 30 seconds apart is approximately 1.6%.

b. $P(t \leq 10) = \int_{-\infty}^{10} e(t)dt = \int_0^{10} 0.2e^{-0.2t}dt = \int_0^{10} 0.2e^{-0.2t}dt = \left(-e^{-0.2t}\right)\Big|_0^{10} = -e^{-2} + e^0 \approx 0.865$

The probability that successive arrivals are 10 seconds or less apart is approximately 86.5%.

c. $P(t>15) = \int_{15}^{\infty} e(t)dt = \lim_{N\to\infty} \int_{15}^{N} 0.2e^{-0.2t}dt = \lim_{N\to\infty} \left(-e^{-0.2t}\Big|_{15}^{N}\right)$

$= \lim_{N\to\infty} \left(-e^{-0.2N} + e^{-3}\right) = 0 + e^{-3} \approx 0.050$

The probability that successive arrivals are more than 15 seconds apart is approximately 5%.

11. a. $\mu = \int_{-\infty}^{\infty} tP(t)dt = \int_0^4 \frac{3}{32}(4t^2 - t^3)dt = \frac{3}{32}\left(\frac{4}{3}t^3 - \frac{1}{4}t^4\right)\Big|_0^4 = \frac{3}{32}\left(\frac{256}{3} - 64\right) = 2$

The mean time is 2 minutes.

b. $\sigma^2 = \int_{-\infty}^{\infty} (t-2)^2 P(t)dt = \int_0^4 \frac{3}{32}(t^2 - 4t + 4)(4t - t^2)dt = \frac{3}{32}\int_0^4 (-t^4 + 8t^4 - 20t^2 + 16t)dt$

$= \frac{3}{32}\left(-\frac{1}{5}t^5 + 2t^4 - \frac{20}{3}t^3 + 8t^2\right)\Big|_0^4 = \frac{3}{32}\left(-\frac{1024}{5} + 512 - \frac{1280}{3} + 128\right)$

$= \frac{3}{32}\left(\frac{128}{15}\right) = 0.8\sigma = \sqrt{0.8} \approx 0.894$

The standard deviation is approximately 0.894 minute.

c. $P(0 \le t \le 1.5) = \int_0^{1.5} P(t)dt = \int_0^{1.5} \frac{3}{32}(4t - t^2)dt = \frac{3}{32}\left(\frac{4}{2}t^2 - \frac{1}{3}t^3\right)\Big|_0^4 = \frac{3}{32}(4.5 - 1.125) \approx 0.316$

The likelihood that any child between the ages of 8 and 10 learns the rules of the board game in 1.5 minutes or less is approximately 31.6%

d. $P(t \ge 3) = \int_3^{\infty} P(t)dt = \int_3^4 \frac{3}{32}(4t - t^2)dt = \frac{3}{32}\left(\frac{4}{2}t^2 - \frac{1}{3}t^3\right)\Big|_3^4 = \frac{3}{32}\left(32 - \frac{64}{3} - 18 + 9\right) \approx 0.156$

There is approximately 15.6% chance that any child between the ages of 8 and 10 takes 3 or more minutes to learn the rules of the board game.

13. a. The number of customers who require daily ATM service is increasing the fastest at one standard deviation less than the mean, 167 – 30 = 137 customers.

b. It would help them to determine when to perform service on the ATM. It might also help them to estimate how much money needs to be in the ATM at specific times.

c. Use technology to approximate the values of the definite integrals.

i. $P(150 \le x \le 200) = \int_{150}^{200} \frac{1}{30\sqrt{2\pi}} e^{\frac{-(x-167)^2}{2(30^2)}} dx$

≈ 0.579

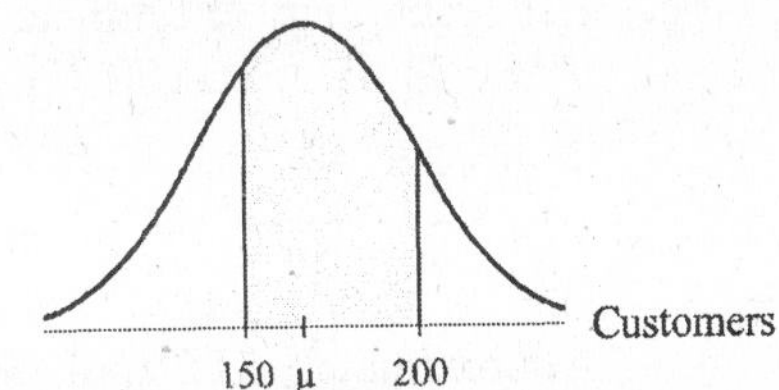

ii. $P(x < 220) = P(x \le \mu) + P(\mu < x < 220)$

$= 0.5 + \int_{167}^{200} \frac{1}{30\sqrt{2\pi}} e^{\frac{-(x-167)^2}{2(30^2)}} dx$

$\approx 0.5 + 0.461 = 0.961$

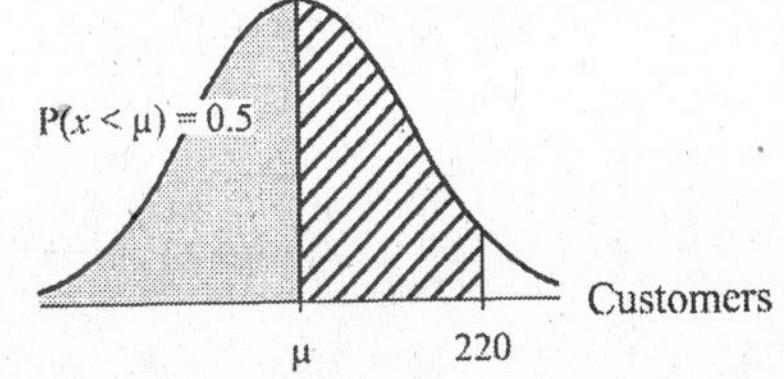

iii. $P(x > 235) = P(x \ge \mu) - P(\mu < x < 235)$

$= 0.5 - \int_{167}^{235} \frac{1}{30\sqrt{2\pi}} e^{\frac{-(x-167)^2}{2(30^2)}} dx$

$\approx 0.5 - 0.49 = 0.01$

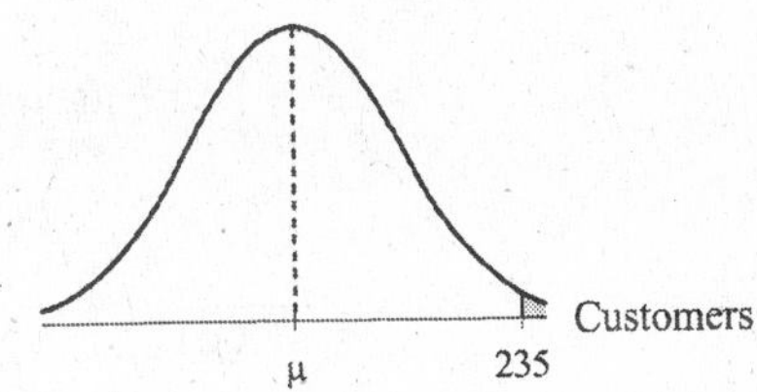

15. a. $6\sigma = 800 - 200 = 600$, so $\sigma = 100$.

b. The realigned mean score is more for each distribution, because the recentering puts the mean of each distribution at 500.

c. 50% of students were expected to make a math score of at least 475 under the original score system because 475 was the mean before the recentering.

d. No, there is not enough information because we do not know the standard deviation.

e. Yes, the shape of the distribution changes in order to make comparison between the math and verbal scores. Note that there is no shift since the overall scale remains from 200 to 800.

f. The scale was recentered for interpretation purposes. The recentering does not reflect any change in student performance. Entrance requirements and other comparisons will now be made on the new scale.

17. a. Use technology to estimate the value of the integral.

$$P(60 \le x \le 80) = \int_{60}^{80} \frac{1}{28.65\sqrt{2\pi}} e^{\frac{-(x-72.3)^2}{2(28.65^2)}} dx \approx 0.272$$

Approximately 27.2% of the students are likely to make a score between 60 and 80.

b. Use technology to estimate the value of the integral.

$$P(x > 90) = 0.5 - \int_{72.3}^{90} \frac{1}{28.65\sqrt{2\pi}} e^{\frac{-(x-72.3)^2}{2(28.65^2)}} dx \approx 0.5 - 0.232 = 0.268$$

Approximately 26.8% of the students are likely to make a score of at least 90.

c. Use technology to estimate the value of the integral.
$P(x < 43.65 \text{ or } x > 100.95) = 1 - P(43.65 \le x \le 100.95)$

$$= 1 - \int_{43.65}^{100.95} \frac{1}{28.65\sqrt{2\pi}} e^{\frac{-(x-72.3)^2}{2(28.65^2)}} dx \approx 1 - 0.683 = 0.317$$

Approximately 31.7% of the students made a score that was more than one standard deviation away from the mean.

d. The rate of change is a maximum at one standard deviation less than the mean, $72.3 - 28.65 = 43.65$.

19. a. $$\mu = \int_{-\infty}^{\infty} xu(x)dx = \frac{1}{b-a}\int_a^b x dx = \frac{1}{b-a}\left(\frac{x^2}{2}\Big|_a^b\right) = \frac{1}{b-a}\left(\frac{b^2-a^2}{2}\right) = \frac{b+a}{2}$$

b. $$\sigma^2 = \frac{1}{b-a}\int_a^b (x-\mu)^2 dx = \frac{1}{b-a}\left[\frac{(x-\mu)^3}{3}\Big|_a^b\right] = \frac{1}{3(b-a)}\left[(b-\mu)^3 - (a-\mu)^3\right]$$

$$= \frac{1}{3(b-a)}\left[\left(b - \frac{b+a}{2}\right)^3 - \left(a - \frac{b+a}{2}\right)^3\right] = \frac{1}{3(b-a)}\left[\left(\frac{b-a}{2}\right)^3 - \left(\frac{a-b}{2}\right)^3\right]$$

$$= \frac{1}{24(b-a)}\left[(b-a)^3-(a-b)^3\right] = \frac{1}{24(b-a)}\left[(b-a)^3+(b-a)^3\right]$$

$$= \frac{2(b-a)^3}{24(b-a)} = \frac{(b-a)^2}{12}$$

So $\sigma = \frac{b-a}{\sqrt{12}}$

c. When $x<a$, $F(x)=\int_{-\infty}^{x} f(t)dt = \int_{-\infty}^{x} 0dt = 0$

When $a \le x \le b$, $F(x)=\int_{-\infty}^{x} f(t)dt = \int_{a}^{x} \frac{1}{b-a}dt = \frac{1}{b-a}\left(t\Big|_a^x\right) = \frac{1}{b-a}(x-a) = \frac{x-a}{b-a}$

When $x>b$, $F(x)=\int_{-\infty}^{x} f(t)dt = \int_{a}^{b} \frac{1}{b-a}dt = \frac{1}{b-a}\left(t\Big|_a^b\right) = \frac{1}{b-a}(b-a)=1$

21. a.

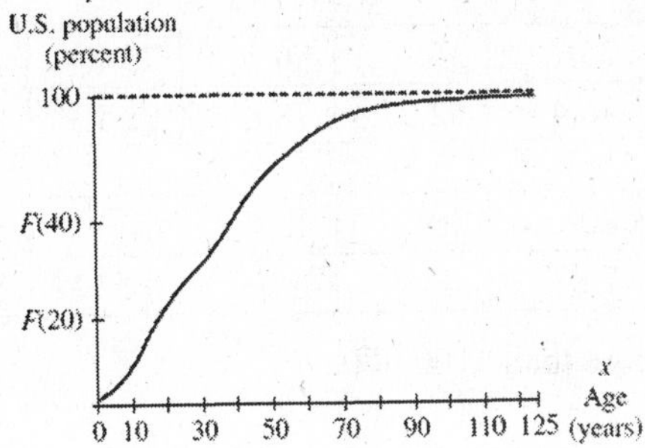

b.

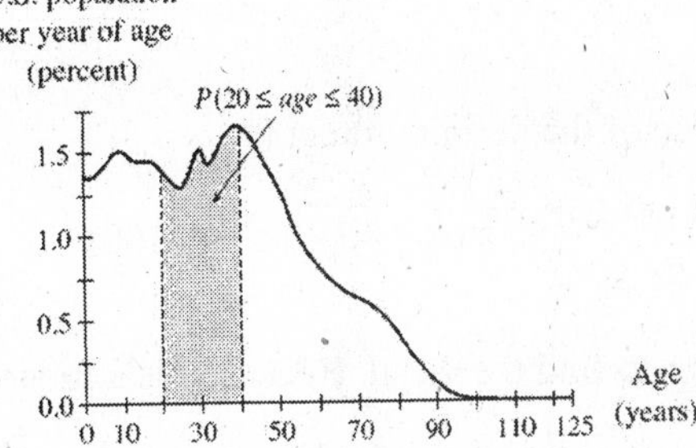

23. a. When $x<0$, $F(x)=0$. When $0 \le x < 1$, $F(x)=\int_{-\infty}^{x} f(t)dt = \int_{0}^{x} 2tdt = t^2\Big|_0^x = x^2$

When $x \ge 1$, $F(x)=1$.

Thus $F(x)=\begin{cases} 0 & \text{when } x<0 \\ x^2 & \text{when } 0 \le x < 15 \\ 1 & \text{when } x \ge 1 \end{cases}$

b. $P(x < 0.67) = \int_{-\infty}^{0.67} f(x)dx = \int_{0}^{0.67} 2xdx = x^2\Big|_0^{0.67} = 0.67^2 = 0.4489$

$P(x < 0.67) = F(0.67) = 0.67^2 = 0.4489$

c. $P(x > 0.25) = 1 - P(x \le 0.25) = 1 - F(0.25) = 1 - 0.25^2 = 0.9375$

d. 

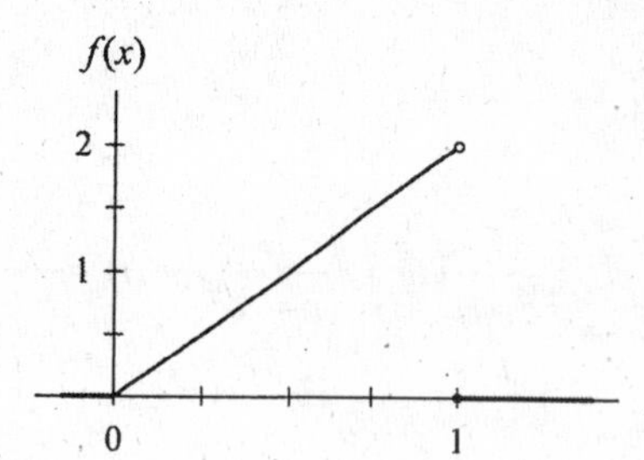

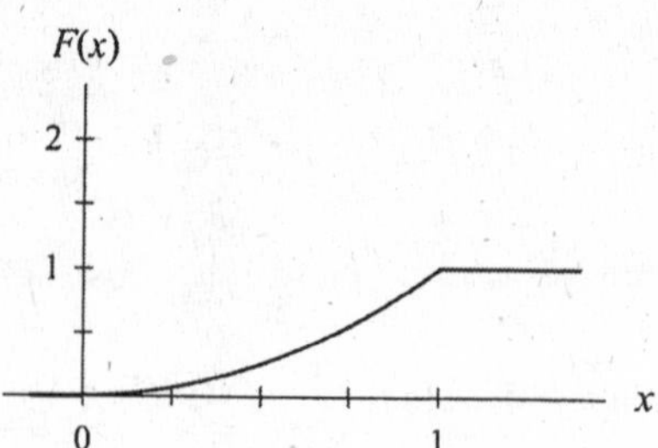

25. a.

x	5	10	15	25	35	50	75	100
$F(x)$	8.2	17.8	27.6	46.4	62.2	78.8	92.1	96.4

x	2.5	7.5	12.5	20	30	42.5	62.5	87.5
$f(x)$	1.64	1.92	1.96	1.88	1.58	1.107	0.532	0.172

b. There are some people who make more than $100,000.

c. Yes, there appears to be an inflection point near $x = 40$, and the limiting value as x increases appears to be 0.

d. The fit appears to be good.

e. Use technology to estimate the value of the definite integral.

$$\int_0^{100} f(x)dx = \int_0^{100} 0.85403395x^{0.56}e^{-0.043798079x}dx \approx 96.41\% \approx F(100)$$

f. Note that we evaluate $\int_0^{100} \frac{xf(x)}{100}dx$ to find the mean. Because f is a function that gives a percent, we divide by 100 to convert it to a probability density function. Use technology to approximate the value of the integral.

$$\mu = \int_0^{100} \frac{xf(x)}{100}dx = \int_0^{100} 0.85403395x^{1.56}e^{-0.043798079x}dx \approx 31.117$$

The integral is in thousands of dollars. Thus the mean is approximately $31,000.

27. a. The distribution on the left (the bell-shaped curve) is characteristic of a normal breeding population. The graph on the right (with three peaks) is what would be expected from a controlled population in which compys were introduced in three batches.

b. The dinosaurs were supposed to be all female and, therefore, not to reproduce, so Malcolm saw the normal curve.

c. Neither graph is a probability density function because the area under each curve is greater than 1.

Chapter 6 Concept Review

1. a. $R(t) = 0.10(3000(12) + 500t) = 0.10(36,000 + 500t) = 3600 + 50t$ dollars per year after t years

   b. Use technology to estimate the value of the integral:

   Six-year future value $= \int_0^T R(t)e^{r(T-t)}dt = \int_0^6 (3600 + 50t)e^{0.083(6-t)}dt = \$29,064$

   c. Present value: $Pe^{rt} = \$29,064$

   $$Pe^{(0.083)(6)} = 29,064$$

   $$P = \frac{29,064}{e^{(0.083)(6)}} \approx \$17,664$$

2. a. The monthly rate of investment is $R(t) = 0.07[3100(1.0004^t)] = 217(1.0004^t)$ dollars per month after t months. There are 8(12)=96 compounding periods.

   96-month future value $= \sum_{t=0}^{95} 217(1.0004^t)\left(1 + \frac{0.082}{12}\right)^{96-t} = \$35,143.80$

   b. Present value: $P\left(1 + \frac{r}{n}\right)^{nt} = \$35,143.80$

   $$P\left(1 + \frac{0.082}{12}\right)^{12(8)} = 35,143.80$$

   $$P = \frac{35,143.80}{\left(1 + \frac{0.082}{12}\right)^{96}} = \$18,277.65$$

3. a. $10,000(0.63^{20}) \approx 1$ fox

   b. $500\left(0.63^{20-t}\right)$ foxes born t years after 1990 that will still be alive in 2010

   c. $10,000(0.63^{20}) + \int_0^{20} 500\left(0.63^{20-t}\right)dt \approx 1083$ foxes

4. a. By solving $D(p) = 0$ we find that the demand curve crosses the p-axis at $p \approx 21.42$ hundred dollars. By solving $D(p) = 30$ we find that the price associated with a demand of 30 fountains is $p \approx 20.94$ hundred dollars. Consumers' willingness and ability to spend is calculated as

   (30 fountains)(20.94 hundred dollars per fountain) + $\int_{20.94}^{21.42} D(p)dp$

   $\approx \$62,827 + \$716 = \$63,543$

   b. At \$1000 each, suppliers will supply $S(10) = 0.3(100) + 8.1(10) + 300 = 411$ fountains. At \$1000 each, the demand will be $D(10) = -100 - 20.6(10) + 900 = 594$ fountains. Supply will not exceed demand at this price.

c. We find the equilibrium point by finding where $S(p) = D(p)$. The input of the equilibrium point is $p \approx 13.11$ hundred dollars. The total social gain is

$$\int_2^{13.11} S(p)dp + \int_{13.11}^{21.42} D(p)dp$$

$$= 4239.66 + 1996.35 \approx 6236.0 \text{ hundred dollars} = \$623{,}600$$

5. a. $P(x < 3.8) = \int_0^{3.8} 0.125x dx = 0.0625x^2 \Big|_0^{3.8} = 0.9025$. The value of x will be smaller than 3.8 approximately 90% of the time. This event is likely to occur.

b.

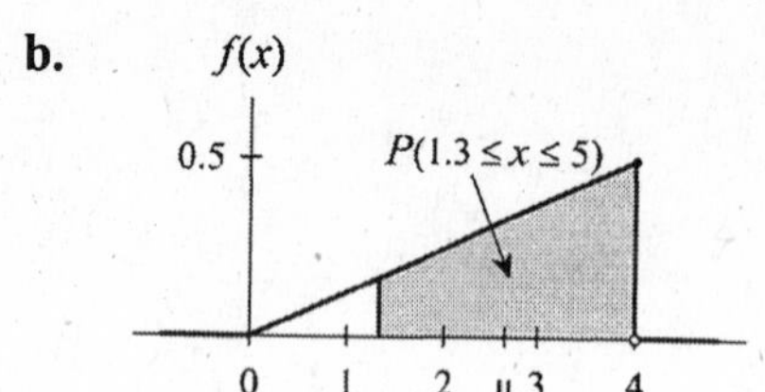

$$P(1.3 \le x \le 5) = \int_{1.3}^{4} 0.125x dx$$

$$= 0.0625x^2 \Big|_{1.3}^{4}$$

$$= 0.894375$$

c. $\mu = \int_0^4 x(0.125x)dx = \frac{0.125}{3} x^3 \Big|_0^4 \approx 2.6667$

d.

$$F(x) = \begin{cases} 0 & \text{when } x < 0 \\ 0.0625x^2 & \text{when } 0 \le x \le 4 \\ 1 & \text{when } x > 4 \end{cases}$$

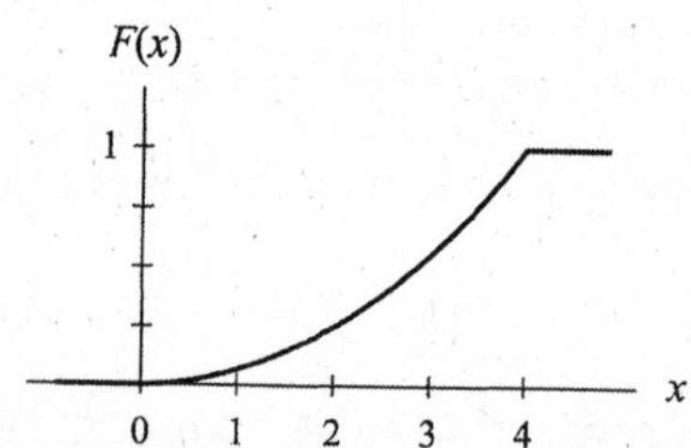

e. $P(1.3 < x < 5) = F(5) - F(1.3) = 1 - 0.0635(1.3^2) = 1 - 0.105625 = 0.894375$

Chapter 7
Repetitive Change: Cycles and Trigonometry

Section 7.1 Functions of Angles: Sine and Cosine

1.

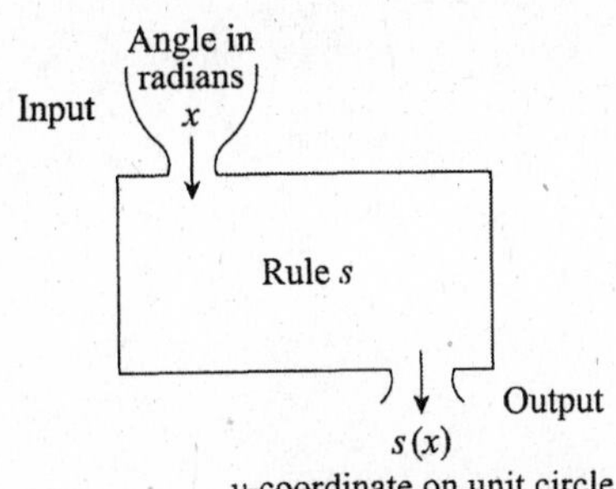

3. a. The maximum value is 1 and occurs at $x = \frac{\pi}{2}$.

   b. The absolute maximum value is 1 and occurs at $x = \frac{\pi}{2} + 2k\pi$, where k is an integer.

   c. The minimum value is -1 and occurs at $\frac{3\pi}{2}$.

   d. The absolute minimum value is -1 and occurs at $x = \frac{3\pi}{2} + 2k\pi$, where k is an integer.

5. a.

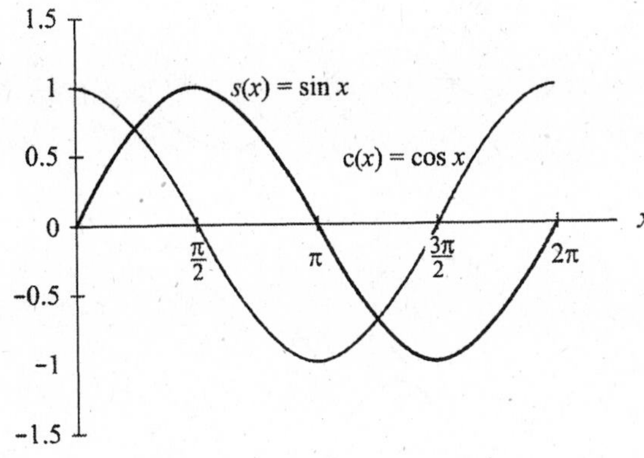

   b. $c(x) = s(x) \approx 0.707$ when $x \approx 0.785$, and $c(x) = s(x) \approx -0.707$ when $x \approx 3.927$.

   c. $x \approx 0.785 + 2k\pi$, where k is an integer, and $x \approx 3.927 + 2k\pi$, where k is an integer.

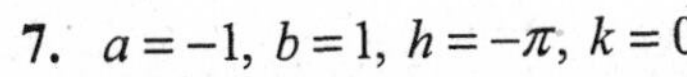

7. $a = -1,\ b = 1,\ h = -\pi,\ k = 0$

   Amplitude: $|a| = 1$ Period: $\frac{2\pi}{b} = \frac{2\pi}{1} = 2\pi$

   Vertical shift: none, $k = 0$

   Horizontal shift: $\frac{|h|}{b} = \frac{|-\pi|}{1} = \pi$; right, $-\pi < 0$

   Horizontal axis reflection: yes, $a < 0$

   At $x = \frac{-h}{b} = \frac{\pi}{1} = \pi$, $g(x)$ is decreasing

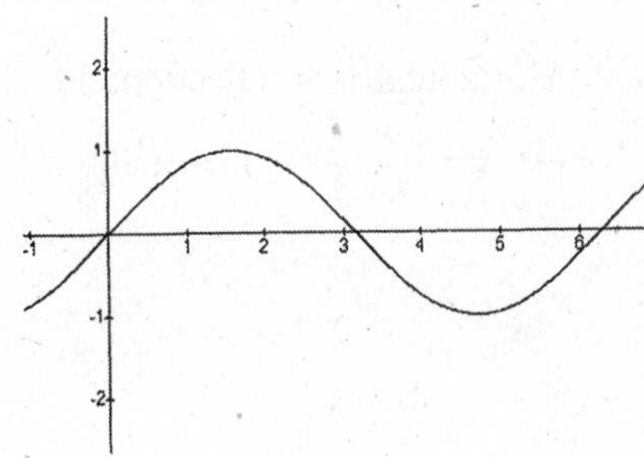

9. Because b is negative in $p(x) = 235\sin(-300x + 100) - 65$,
rewrite $p(x)$ as $p(x) = -235\sin(300x - 100) - 65$
Then, $a = -235,\ b = 300,\ h = -100,\ k = -65$

Amplitude: $|a| = 235$ Period: $\dfrac{2\pi}{300} \approx 0.021$

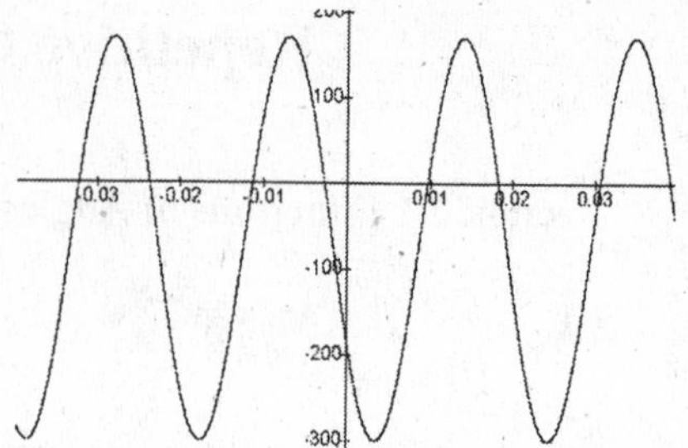

Vertical shift: down 65, $k = -65$

Horizontal shift: $\dfrac{|h|}{b} = \dfrac{|-100|}{300} = -\dfrac{1}{3}$; right, $-100 < 0$

Horizontal axis reflection: yes, $a < 0$

At $x = \dfrac{-h}{b} = \dfrac{100}{300} = \dfrac{1}{3}$, $p(x)$ is decreasing

11. Observe that for a graph of the form $y = a \sin x + k$, where a is positive, the maximum is $k + a$ and the minimum is $k - a$. Use this information to match the graph with its equation.

a. maximum is 11, minimum is 5, match is iii
b. maximum is 0, minimum is -6, match is vi
c. maximum is 5, minimum is -1, match is i
d. maximum is 12, minimum is 4, match is v
e. maximum is -1, minimum is -5, match is ii
f. maximum is 4, minimum is 0, match is iv

13. Amplitude: 1 Period: $\dfrac{2\pi}{2} = \pi$

Vertical shift: up 3
Horizontal shift: none
Horizontal axis reflection: no

At $x = \dfrac{-h}{b} = 0$: increasing

15. Amplitude: 2 Period: $\dfrac{2\pi}{1} = 2\pi$

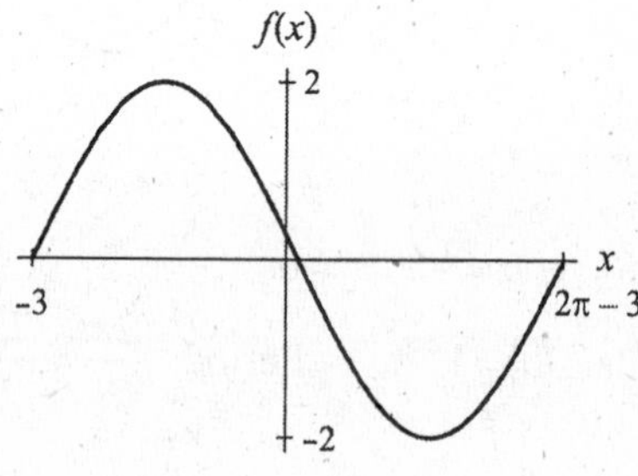

Vertical shift: none

Horizontal shift: left $\dfrac{3}{1} = 3$

Horizontal axis reflection: no

At $x = \dfrac{-h}{b} = -3$: increasing

17. Amplitude: 2 Period: $\frac{2\pi}{2} = \pi$

Vertical shift: none

Horizontal shift: right $\frac{3}{2}$

Horizontal axis reflection: yes

At $x = \frac{3}{2}$: decreasing

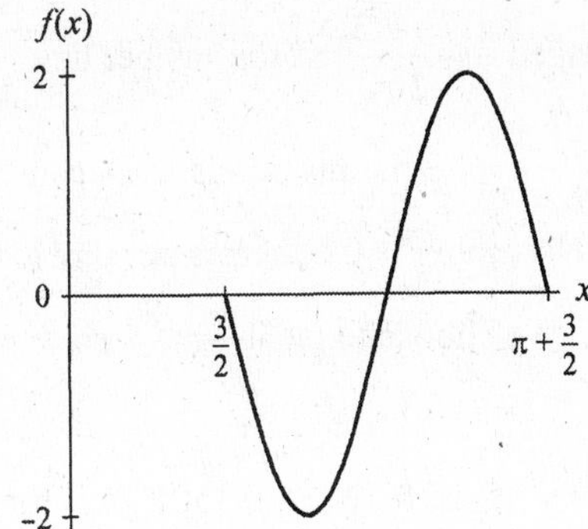

19. The members of each pair of graphs are identical. Generalization: Begin with a sine graph that is not vertically shifted, and reflect it across the horizontal axis. Begin with the same sine graph and shift it right or left by an integer multiple of half the period of the original graph. The new graph will be identical to the reflected graph.

21. a. The graph appears to have been rotated about the horizontal axis.

b. The graph appears not to have rotated at all. The graph is symmetric about the vertical axis.

c. $f(x) = \cos x$ is equivalent to $f(x) = \cos(-x)$ for all values of x. $f(x) = \cos x$ is the additive inverse of $f(x) = -\cos x$ for all values of x.

23. If we are given a graph of a sine function, the period is the difference in input between a min output value and the next succeeding minimum output value (or *the difference in input between a maximum output value and the next succeeding maximum output value).* We determine the horizontal shift $\frac{|h|}{b}$ of a sine function from its graph by determining the input value for which the output value is the average value *d*. If the shift is to the right, *h* is negative. If the shift is to the left, *h* is positive. If we are given data that can be modeled by a sine function, the amplitude is half the difference between the maximum and minimum output data values. If the amplitude is less than the maximum output value of the data, the vertical shift is equal to the difference in the maximum output value and the amplitude.

Section 7.2 Sine Functions as Models

1. a. The data are cyclic because the residential gas usage is determined by the daily temperature, which usually repeats itself over a year's time. We see this cycle as the second year of November, December, and January begins.

b. Amplitude $\approx \frac{\text{max} - \text{min}}{2} = \frac{3.3 - 0.3}{2} = 1.5$ therms per day, $a = 1.5$; the maximum value of the function = 3.3 which is the sum $a + k$, so $k = 1.8$.

c. Period $\approx$ 12 months, period $= \frac{2\pi}{b}$, so $12 = \frac{2\pi}{b}$ and ; $b = \frac{2\pi}{12} \approx 0.524$

Now we have $f(x) = 1.5\sin(0.524x + h) + 1.8$. We need to find the horizontal shift. If the first January is month 1, then the first December is month 0. The gas usage in December averaged $\frac{1.7+2.3}{2} = 2$ therms per day. The gas usage in November averaged $\frac{1.2+1.4}{2} = 1.3$ therms per day. December is 0.2 therms per day higher than the average of 1.8, November is 0.5 therms per day lower. We want a shift approximately 2/7 to the left of 0. We choose to let $\frac{|h|}{\pi/6} = \frac{2}{7}$, so $h \approx 0.150$, recalling that $h > 0$ indicates a shift to the left.

d. $G(x) = 1.5\sin(0.524x + 0.150) + 1.8$ therms per day is the average daily gas usage by a residential customer in Reno, Nevada in month x, where x is the month of the year, x = 1 in January.

3. a.

Daily mean temperature (°F)

90

10

Month

1 (January)

12 (December)

The data are cyclic because temperatures vary as a result of seasons, which are cyclic. The average daily temperature will repeat itself every year.

b. Amplitude $= \frac{76.9 - 21.1}{2} = 27.9$ °F; Vertical shift = 21.1 + 27.9 = 49°F

c. Period ≈ 12 months

To estimate the horizontal shift, we find the value closest to the vertical shift of 49°F. This value is 51.9°F which occurs in April. Thus the data are shifted right by 4 months:

Horizontal shift ≈ 4 months to the right, $\frac{|h|}{b} = 4$

d. On the basis of the answers to parts b and c, we have $a = 27.9$, $b = \frac{2\pi}{12} = \frac{\pi}{6} \approx 0.52$, $h = 4\left(\frac{\pi}{6}\right) = \frac{2\pi}{3} \approx 2.09$, and $k = 49$.

$m(x) = 27.9\sin\left(\frac{\pi}{6}x - \frac{2\pi}{3}\right) + 49 \approx 27.9\sin(0.52x - 2.09) + 49$ °F

gives the mean daily temperature during the xth month of the year

e.

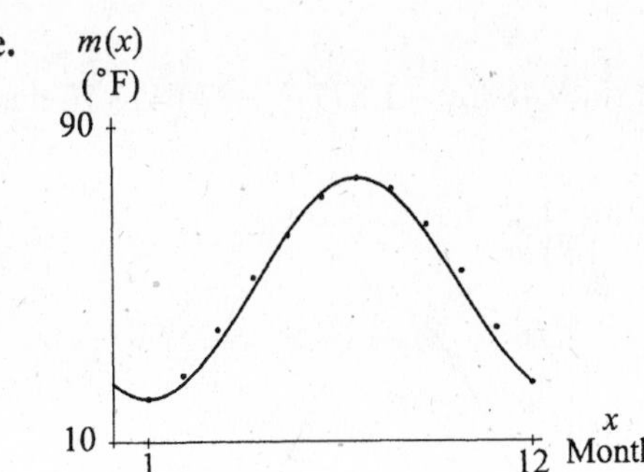

The fit appears to be reasonable.

f. $x = 7$ in July and $f(7) = 27.9$

$\sin(0.52 \cdot 7 - 2.09) + 49 \approx 76.9$.

The mean daily temperature in July of this year is approximately 76.9 °F. This result agrees with the value in the table. Although the function is an excellent description of the normal average daily temperature for the years 1961-1990, it cannot predict with any certainty the mean daily temperatures in Omaha for a particular July.

5. a. The low temperatures should have the same period and approximately the same amplitude and horizontal shift as the average temperatures, but they should have a smaller vertical shift.

b. Amplitude $\approx \dfrac{65.9 - 10.9}{2} = 27.5$ °F; Vertical shift $\approx 10.9 + 27.5 = 38.4$°F

Horizontal shift $\approx$ 4 months to the right; Period $\approx$ 12 months

As expected, the period and horizontal shift are the same. The amplitude is slightly lower than for the mean temperature data, and the vertical shift is less.

c. We calculate b using the equation $12 = \frac{2\pi}{b}$, so $b = \frac{\pi}{6}$. We calculate h using the equation $\frac{|h|}{b} = 4$ and the information that the horizontal shift is to the right, so h is negative: $h = -4b$ $= \frac{-2\pi}{3}$. Using $a = 27.5$ and $k = 38.4$, we have $f(x) = 27.5\sin(\frac{\pi}{6}x - \frac{2\pi}{3}) + 38.4$ °F is the normal daily low temperature in Omaha during the xth month of the year.

7. a. A possible model is $G(x) = 1.610\sin(0.534x - 0.322) + 1.591$ therms per day is the gas usage for a residential customer where x is the number of months after November 2000.

Amplitude: $a = 1.610$ therms per day

Period $\approx \dfrac{2\pi}{0.534} \approx 11.8$ months Horizontal shift $= \dfrac{0.322}{0.534} \approx 0.603$, $.603 \cdot 12 \approx 7.2$ month2 to the right

Vertical shift $\approx$ up 1.591 therms per day

b. The expected value is approximately 1.6 therms per day.

c. The average gas bill is approximately
(1.591 therms per day)(30 days)($0.56532 per therm) $\approx$ $26.89.

9. Using technology, a model is $g(x) = 28.5328\sin(0.4789x - 1.8013) + 47.9847$ °F during the xth month of the year. The b and k values in Activity 2 are slightly more than those for the technology model, while a and h are slightly less than those for the technology model. The function obtained by using technology seems to fit the data better than the model found in Activity 3.

11. a. Using technology, a model is $g(x) = 27.3572 \sin(0.4886x - 1.8777) + 37.6413$ °F during the xth month of the year.

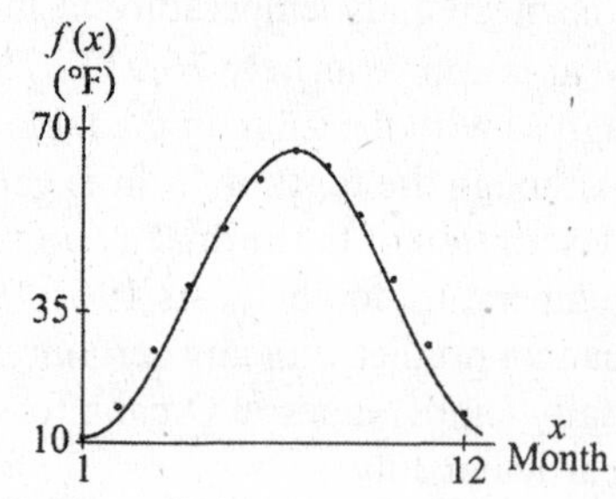

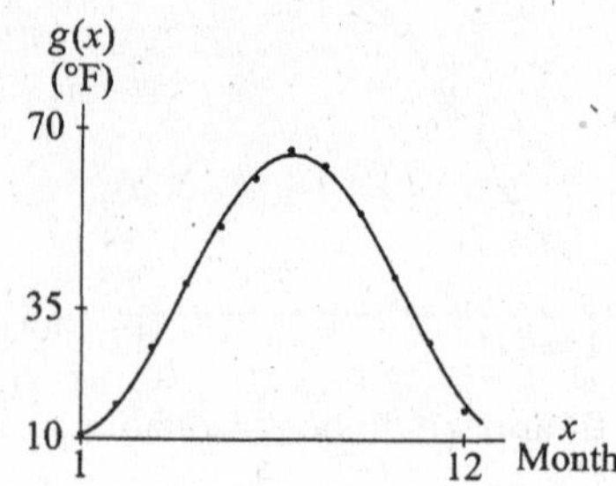

Both models appear to fit the data well.

b. *One possible answer*: The model generated by technology takes less algebraic work to obtain.

13. a. $M(x) = 0.932\sin(0.467x - 2.940) + 8.737$ billion trips on mass transit each year, x years after 1992

b. $T(x) = -0.0052x^3 + 0.127x^2 - 0.573x + 8.56$ billion trips x years after 1992

c. The cubic function fits the data fairly well, but the sine function better follows the curvature of the data and provides a very good fit.

d. $M(14) \approx 8.32$ billion trips

$T(14) \approx 11.24$ billion trips

$T(x)$ seems a better model because it reflects higher costs to run individual automobiles and a growing population. These factors may increase the use of mass transportation. However, answers may vary.

15. a. The highest mean daily temperature at Fairbanks, Alaska is $a + k = 37 + 25 = 62$ °F. The lowest mean daily temperature is $-a + k = -37 + 25 = -12$ °F.

b. The average of the highest and lowest mean daily temperatures is $\dfrac{62 + (-12)}{2} = 25$°F, which is k, the vertical shift of the graph of f.

17. a. Let x be the number of years since 1949. The data appear concave up between 1949 and 1955, concave down between 1956 and 1960, and concave up between 1961 and 1963. A cubic model is not a good fit because more than one inflection point are indicated by the data. Possibly a piecewise continuous model that is quadratic between 1949 and 1955 and cubic between 1956 and 1963 would fit the data. However, the sine model seems most appropriate.

b. Using technology, a sine model is $f(x) = 41.5473 \sin(0.5284x - 2.9132) + 194.4930$ aircraft x years after 1949.

c. 1964 is 15 years since 1949. $f(15) \approx 155$ aircraft

d. There are more factors involved in production than simple replacement of a fleet of airplanes. The sine model may not be a good model for continued extrapolation.

19. a.

Hours	°F	Hours	°F
5	37	65	60
11	44.5	71	47
17	52	77	34
23	47	83	45
29	42	89	56
35	49	95	46.5
41	56	101	37
47	49	107	45.5
53	42	113	54
59	51		

b. Using technology, a sine model is $f(x) = 8.7094\sin(0.2608x - 2.8404) + 47.0895$ °F x hours after midnight on Wednesday. The period of this model is $\dfrac{2\pi}{0.2608} \approx 24$ hours.

c. High temperature $= k + a \approx 56$°F; low temperature $= k - a \approx 38$°F

d. The greatest discrepancies between the model and the data occur on Friday afternoon and Saturday morning. The hydroelectric plant's energy may also be needed Wednesday afternoon and Sunday morning.

21. Electricity is produced on demand, not stored until the demand occurs. Locally dependent answers will vary.

23. Without using technology:

Vertical shift $= k = \dfrac{\text{max} + \text{min}}{2} = \dfrac{18.5 + 4.5}{2} = 11.5$

Amplitude $= a = \dfrac{18.5 - 4.5}{2} = 7$

Period $= 365 = \dfrac{2\pi}{b}$, so $b = \dfrac{2\pi}{365}$

Horizontal shift $= \dfrac{181 + 361}{2} = 271$ to the right

$\dfrac{|h|}{b} = 271$, so $|h| = 271b = \dfrac{271(2\pi)}{365}$ Because the shift is to the right, h is negative.

The model is $D(t) = 7\sin\left(\dfrac{2\pi}{365}t - \dfrac{542\pi}{365}\right) + 11.5$ hours of daylight t days after December 31 of the previous year.

Using technology, the model is $d(t) = 6.6767\sin(0.0164t - 1.9083) + 11.7299$ hours of daylight t days after December 31 of the previous year.

25. Excel Activity

Use technology to find sine models for each odor. For each model, you may need to give an approximate period to avoid a singular matrix in creating a sine model. To estimate the period, determine when the firing rate first returns to its beginning value. For example, for cherry odors the starting value is 14 and the value is 15 at 50 m/s, so we use 50 as an approximate period. Models may vary.

Cherry odor (approximate a period of 50 ms):
$c(x) = 6.9302 \sin(0.1129x + 2.0378) + 8.8556$ firings/ms with period $\approx$ 55.7 ms

Apple odor (approximate a period of 40 ms):
$a(x) = 7.0726 \sin(0.1327x + 1.6205) + 10.5255$ firings/ms with period $\approx$ 47.3 ms

Cherry/apple odor (approximate a period of 50 ms):
$t(x) = 3.2213 \sin(0.1203x + 1.9576) + 5.6179$ firings/ms with period $\approx$ 52.2 ms

The periods appear to be close, although we do not have enough information to determine if the difference between them is significant or not.

27. Excel Activity

a. Amplitude $\approx \dfrac{33-0}{2} = 16.5$ thousand lizards; period $\approx$ 12 months

vertical shift $\approx$ 16.5 thousand lizards

b. $L(m) = 16.298\sin(0.537m - 1.801) + 11.206$ thousand lizards in month m where m is the number of months since the beginning of year 1. The amplitude of this model is approximately 16.3 thousand lizards, only 0.2 thousand lizards less than the estimated amplitude in part *a*. The period is $\dfrac{2\pi}{0.537} \approx 11.7$ months, compared with the 12 months estimated in part *a*. The vertical shift is significantly less than the one estimated in part *a*.

c.

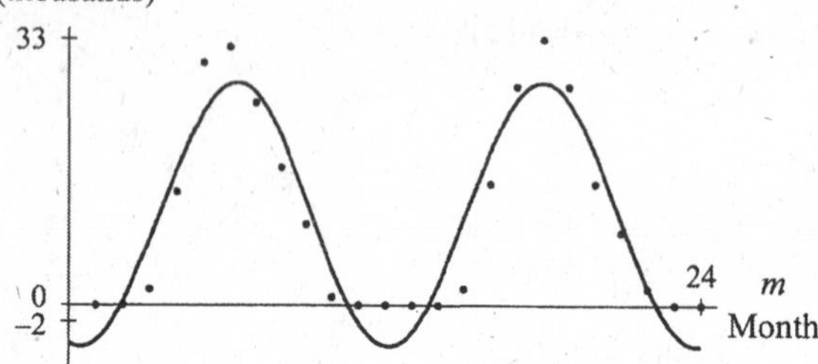

The graph appears to match the period of the data well, but the graph falls below the horizontal axis, and doesn't reach the highest values.

d. $l(x) = 15.388\sin(0.787x - 1.667) + 16.346$ thousand lizards in month x where $x = 1$ through $x = 8$ correspond to March through October of year 1 and $x = 9$ through $x = 16$ correspond to March through October of year 2. The amplitude of this function is 15.388, approximately 1 less than the amplitude of the equation in part *b*. The period is approximately 8 months instead of the 11.9 months in part *b*. This smaller period is the result of deleting the data for 4 months of each year. The vertical shift of 16.3 is greater than the vertical shift of the equation in part *b*. This vertical shift better matches the shift calculated in part *a*. This equation fits the modified data well.

Section 7.3 Rates of Change and Derivatives

1. $f'(x) = \frac{d}{dx}(\sin 3x) + 0 = 3\cos 3x$

3. $\frac{dt}{dr} = 5.2\frac{d}{dr}\cos(0.45r + \pi) + 80 - 0 = 5.2(-0.45)\sin(0.45r + \pi) + 80$

5. $h'(x) = 2.08\frac{d}{dx}\sin(-0.16x + 12.3) + 3.58\frac{d}{dx}\cos(2.7x + 8.1) - 0$

$= 2.08(-0.16)\cos(-0.16x + 12.3) + 3.58(2.7)[-\sin(2.7x + 8.1)]$

$= -0.3328\cos(-0.16x + 12.3) - 9.666\sin(2.7x + 8.1)$

$h''(x) = -0.3328\frac{d}{dx}\cos(-0.16x + 12.3) - 9.666\frac{d}{dx}\sin(2.7x + 8.1)$

$= -0.3328(-0.16)[-\sin(-0.16x + 12.3)] - 9.666(2.7)\cos(2.7x + 8.1)$

$= -0.053248\sin(-0.16x + 12.3) - 26.0982\cos(2.7x + 8.1)$

7. a. It takes the satellite 138 – 18 = 120 minutes = 2 hours to complete one orbit.

b. The greatest distance south of the equator is 3050 kilometers at 48 minutes after launching.

c.

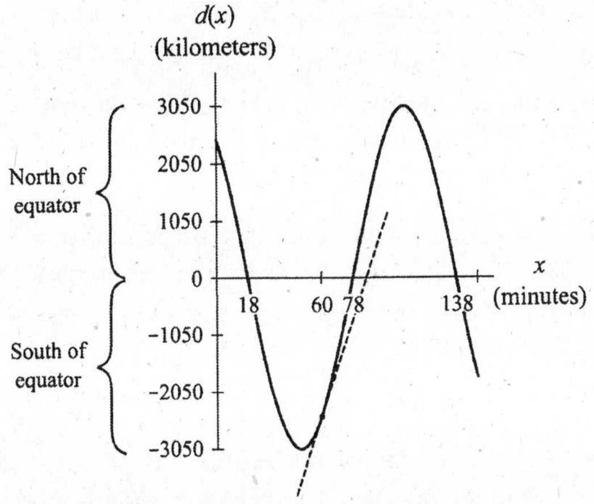

d. The rate of change is approximately 93.9 kilometers per minute.

e. No, the answer to part *d* is the rate of change of the distance from the equator with respect to time. It is not speed, which is the rate of change of the total distance traveled with respect to the traveling time.

9. a. Rate of change in 1992 ≈ -0.4 billion trips per year
Rate of change in 1996 ≈ 0.2 billion trips per year
Rate of change in 2000 ≈ 0.3 billion trips per year

b. Approximately 9.7 billion trips

11. a. *One possible answer:*
Approximately $9 - 17 = -8$ mm/day

b. Percentage change $= \frac{9-17}{17} 100\% \approx 47.1\%$ decline

Average rate of change $= \frac{9-17}{11-6} \approx -1.6$ mm/day/month

c. *One possible answer:* Approximately 2.4 mm per day per month. Extraterrestrial radiation in Amarillo is increasing by 2.4 millimeters of equivalent water evaporation per day per month in March.

d. *One possible answer:* Percentage rate of change $= \frac{2.3}{12.5} 100\% \approx 18.4\%$ per month

13. a. $R(x) = 0.7c(x) + 0.2p(x)$

$$= 1.32727\sin(0.0186x + 1.1801) + 1.46052\cos(0.0197x - 3.7526) + 7.90076$$

million dollars, where x is the day of the year

b. $R'(x) = 1.32727(0.0186)\cos(0.0186x + 1.1801) -$

$1.46052(0.0197)\sin(0.0197x - 3.7526)$ million dollars per day

where x is the day of the year

c. $R'(46) \approx -0.003$ million dollars per day
Combined sales were decreasing by approximately \$3000 per day on February 15, 1992.

d. Using technology, we find that $R'(x) = R'(46)$ when $x \approx 220$ and $x \approx 293$. Thus, the revenue was decreasing at the same rate as that found in part *c* at approximately 220 and 293 days after the beginning of 1992. (These values correspond to August 7 and October 19. Recall that 1992 was a leap year.)

e. Using technology, we find that $R'(x) = -R'(46)$ when $x \approx 90$, $x \approx 166$, and $x \approx 339$. Thus the revenue was increasing at the same rate at which it was decreasing in part *c* at approximately 90, 166, and 339 days after the beginning of 1992. (These values correspond to March 30, June 14, and December 4.)

15. $B'(m) = 22.926(0.510)\cos(0.510m - 2.151)$ °F per month at the end of the mth month. The end of November corresponds to $m = 11$ and the middle of March corresponds to $m = 4.5$.

$B'(11) \approx -11.1$°F per month and $B'(4.5) \approx 11.6$°F per month

The normal daily high temperature in Boston is decreasing at the end of November at a rate of approximately 11.11°F per day and is increasing in the middle of March at a rate of approximately 11.57°F per day.

17. a. From 6 m/sec to 10 m/sec: change $= P(10) - P(6) \approx -17.42$ kW

From 10 m/sec to 14 m/sec: change $= P(14) - P(10) \approx -17.26$ kW

From 14 m/sec to 18 m/sec: change $= P(18) - P(14) \approx -0.29$ kW

b. From 6 m/sec to 10 m/sec: Percentage change $= \dfrac{P(10) - P(6)}{P(6)} 100\% = 38.4\%$ decline

Average rate of change $= \dfrac{P(10) - P(6)}{10 - 6} \approx -4.36$ kW/m/sec

From 10 m/sec to 14 m/sec: Percentage change $= \dfrac{P(14) - P(10)}{P(10)} 100\% = 61.7\%$ decline

Average rate of change $= \dfrac{P(14) - P(10)}{14 - 10} \approx -4.31$ kW/m/sec

From 18 m/sec to 14 m/sec: Percentage change $= \dfrac{P(18) - P(14)}{P(14)} 100\% = 2.7\%$ decline

Average rate of change $= \dfrac{P(18) - P(14)}{18 - 14} \approx -0.07$ kW/m/sec

c. $P'(s) = 20.204(0.258)\cos(0.258s + 0.570)$ kW/m/sec when the wind speed is s m/sec

d. $P'(6) \approx -2.71$ kW/m/sec

The power output of the engine is decreasing at the rate of 2.7 kilowatts per meter per second at a wind speed of 6 m/s.

$P'(10) \approx -5.21$ kW/m/sec

When the wind speed is 10 m/s and increases to 11 m/s, the power output of the engine decreases by approximately 5.2 kilowatts

$P'(14) \approx -2.64$ kW/m/sec

When the wind speed is 14 m/s and increases to 15 m/s, the power output of the engine decreases at by approximately 2.6.

$P'(18) \approx 2.51$ kW/m/sec

At a wind speed of 18 m/s, the power output of the engine is increasing at the rate of 2.5 kilowatts per meter per second

e. At 6 meters per second: Percentage rate of change $= \dfrac{P'(6)}{P(6)} 100\% \approx -6.0\%$ per m/sec

At 10 meters per second: Percentage rate of change $= \dfrac{P'(10)}{P(10)} 100\% \approx -18.6\%$ per m/sec

At 14 meters per second: Percentage rate of change $= \dfrac{P'(14)}{P(14)} 100\% \approx -24.7\%$ per m/sec

At 18 meters per second: Percentage rate of change $= \dfrac{P'(18)}{P(18)} 100\% \approx -24.1\%$ per m/sec

19. a. Using technology, a sine model is $d(t) = 60.7407 \sin(0.1206t + 0.3134) + 99.9194$ deaths per 100,000 people per week, where t is the number of weeks since January 1, 1923. (Answers will vary.)

b. $d'(t) = 60.7407(0.1206)\cos(0.1206t + 0.3134)$ deaths per 100,000 people per week per week, where t is the number of weeks since January 1, 1923
The middle of 1924 corresponds to $t = 78$.
$d'(78) \approx -7.01$ deaths per 100,000 people per week per week

The end of 1924 corresponds to $t = 104$.
$d'(104) \approx 7.02$ deaths per 100,000 people per week per week

21. a. $D(d) = 23.677\sin(0.017d - 1.312) - 0.292$ degrees on the dth day of the year
The amplitude of the model is 23.7 degrees. This is the greatest angle of declination the sun reaches. The period is approximately 374 days (calculated with the unrounded b value). We expect the sun's declination to complete a cycle every year. The period in the model is slightly long.

b. The equinoxes occur at $d \approx 78.7$ and $d \approx 264.4$.

$D'(78.7) \approx 0.4$ degrees per day; $D'(264.4) \approx -0.4$ degree per day

At the equinoxes, the declination of the sun is changing by approximately 0.4 degree per day.

c. The summer and winter solstices occur when the declination is greatest in both north and south directions. On the graph, the summer solstice corresponds to the maximum and the winter solstice corresponds to the minimum. At these points, the rate of change is zero.

23. Inside: $u(x) = 2.4x$ Outside: $y(u) = \sin u$
Derivative of inside: $u'(x) = 2.4$ Derivative of outside: $y'(u) = \cos u$

$$y'(x) = u'(x)y'(u(x)) = 2.4\cos(2.4x)$$

25. Inside: $u(x) = \sin x - 7$ Outside: $y(u) = 4u^2 + 8u + 13$
Derivative of inside: $u'(x) = \cos x$ Derivative of outside: $y'(u) = 8u + 8$

$$y'(x) = u'(x)y'(u(x)) = (\cos x)(8u + 8) = (\cos x)[8(\sin x - 7) + 8]$$
$$= 8\cos x(\sin x - 7 + 1) = 8\cos x(\sin x - 6)$$

27. Inside: $u(x) = \sin x$ Derivative of inside: $\cos x$
Outside: $y(u) = \ln u$ Derivative of outside: $y'(u) = \frac{1}{u}$

$$y'(x) = u'(x)y'(u(x)) = (\cos x)\frac{1}{u} = (\cos x)\frac{1}{\sin x} = \frac{\cos x}{\sin x}$$

29. a. Using technology, a sine model is $f(x) = 59.6582 \sin(0.5250x - 1.9621) + 105.2150$ thousands of dollars in sales, where x is the number of months since December of the year before Year 1. The fit is fairly close, except near the maxima and minima for the later years.

b. The average of the high and low sales for Year 1 is $\frac{167+50}{2} = \$108.5$ thousand

The average of the high and low sales for Year 2 is $\frac{174+52}{2} = \$113$ thousand

The average of the high and low sales for Year 3 is $\frac{175+55}{2} = \$115$ thousand

The average of the high and low sales for Year 4 is $\frac{180+58}{2} = \$119$ thousand

c. Enter the following data: (5, 108.5), (17, 113), (29, 115), (41, 119). Using technology, a linear function is $y = 0.2792x + 107.4542$ thousand dollars (average of high and low sales), where x is the number of months since December of the year before Year 1.

d. The new model is $f(x) = 59.6582 \sin(0.5250x - 1.9621) + 0.2792x + 107.4542$ thousands of dollars in sales, where x is the number of months since December of the year before Year 1. The fit is better than that of the function in part *a*.

e. $f'(x) = 59.6582(0.5250) \cos(0.5250x - 1.9621) + 0.2792$ thousands of dollars in sales per month, where x is the number of months since December of the year before Year 1.

September of the third year corresponds to $x = 33$.
$f'(33) \approx -\$29.5$ thousand dollars in sales per month
In September of the third year, the sales of ice cream were decreasing at a rate of approximately $29,500 per month.

January of the fourth year corresponds to $x = 37$.
$f'(37) \approx 6.1$ thousand dollars in sales per month
In January of the third year, the sales of ice cream were increasing at a rate of approximately $5700 per month.

Section 7.4 Extrema and Points of Inflection

1. a. Between $x = 0$ and $x \approx 13.45$, the maximum value of 9.669 occurs at $x \approx 9.66$. The minimum value of 7.805 occurs at $x \approx 2.93$.

b. Between $x = 0$ and $x = 8$, the greatest number of yearly mass transit trips between 1992 and 2000 was approximately 9.4 billion (in 2000). The least number of yearly mass transit trips between 1992 and 2000 was approximately 7.805 billion (in 1995).

3. a. The maximum daily mean temperature is approximately 76.9°F (in July).
The minimum daily mean temperature is approximately 21.1°F (in January).

b. Using technology, a sine model is $t(x) = 28.5328 \sin(0.4789x - 1.8013) + 47.9847$ °F is the daily mean temperature in Omaha, Nebraska, in the xth month of the year.
The derivative is $t'(x) = 28.5328(0.4789) \cos(0.4789x - 1.8013)$ °F per month
For $0 < x \le 12$, $t'(x) = 0$ when $x \approx 0.481$ and $x \approx 7.041$, respectively. The maximum point on the model is (7.041, 76.517) and the minimum point on the model is (0.481, 19.452).

Thus, according to the model, the maximum normal daily mean temperature in Omaha between 1961 and 1990 was approximately 76.5°F and occurred at the beginning of August. The minimum normal daily mean temperature was approximately 19.5°F and occurred near the middle of January.

c. The maximum and minimum temperatures obtained with the model are slightly lower than those found using the data. Because the data are based on a monthly average over a 30-year period, actual high and low temperatures in any given year are likely to vary from those found in parts *a* and *b*.

5. a. One possible answer is 7 – 10, but answers may vary since the intervals for 1 year are small and difficult to read from the graph. Some of the peaks of the sine model that appear to be within 1 year of a peak in the wheat price index include the peak between 1762 and 1770, the peak between 1778 and 1786, and the peak between 1834 and 1842.

b. One possible answer is 9 – 10, but answers may vary since the intervals for 1 year are small and difficult to read from the graph. Some of the valleys of the sine model that appear to be within 1 year of a valley in the wheat price index include the valley near 1770, the valley near 1778, and the valley near 1858.

c. For $t > 113$, the next maximum of 100.2 occurs when $t \approx 120.246$ and the next minimum of -100.2 occurs when $t \approx 116.246$. The next maximum would occur during 1882 and the next minimum would occur during 1878.

d. *One possible answer:* The maxima and minima predicted by the sine curve do not always match up with the index numbers of the price of wheat. Using the sine curve for prediction is risky.

7. a. The blue graph with the summer solstice labeled is the number of hours of daylight. The black graph is the number of hours of darkness.

b. Period $\approx$ 12 months
From the graph, the highest point is approximately 20 hours and the lowest point is approximately 5 hours. Thus amplitude $\approx \frac{20-5}{2} = 7.5$ hours and vertical shift $\approx \frac{20+5}{2} = \approx 12.5$ hours.

c. *One possible answer:* The graph is cyclical and repeats every 12 months.

Month	Hours	Month	Hours
Jan	5	July	19
Feb	8	Aug	17
Mar	10.5	Sept	14.5
Apr	13.5	Oct	11.5
May	16.5	Nov	8.5
June	19	Dec	6

d. Using technology, a sine model is $H(m) = 7.021 \sin(0.4803m - 1.6189) + 11.7855$ hours of daylight at the beginning of the *m*th month.
Period $= \frac{2\pi}{0.4803} \approx 13$ months, amplitude $\approx$ 7.0 hours, vertical shift $\approx$ 11.8 hours

The period in the model found using technology is longer, the amplitude is less, and the vertical ship is less.

9. Answers will vary depending on location.

11. a. $a'(m) = 27.1(0.485)\cos(0.485x - 1.707)$ watts per centimeter squared per month when $m = 1$ in January, $m = 2$ in February, and so on.

b. Between $m = 0$ and $m = 12$, $a'(m) = 0$ when $m \approx 0.2808$ and $m \approx 6.7853$.
$a(0.2808) \approx 5.8$ and $c(6.7853) \approx 60$. We compare these outputs with those at the endpoints: $a(0) \approx 6.1$ and $c(61) \approx 10.5$.
Radiation is at its highest level in July and at its lowest level in January.

c. The highest level of radiation is 60 watts per centimeter squared. The lowest level of radiation is 5.8 watts per centimeter squared.

13. a. $T'(x) = 0.565\cos(0.469x - 2.293)$ percent per year x years after 1988.

b. For one period of T, between $x = 0$ and $x \approx 13.4$, solve the equation $T'(x) = 0$ to find that $x \approx 1.54$ (input for relative minimum) and $x \approx 8.24$ (input for relative maximum). Compare the outputs at these inputs to $T(0)$ and $T(13.4)$ to find the absolute extrema. The smallest after-tax profit rate on investment, approximately 5.18%, occurred in 1990 and the largest rate, approximately 7.5%, occurred in 1996.

$$T''(x) = 0.565\left(\frac{d}{du}\cos(u)\right)\left(\frac{d}{dx}(0.469x - 2.293)\right) \quad \text{where } u = 0.469x - 2.293$$
$$= 0.565(-\sin(0.469x - 2.293))(0.469)$$
$$= 0.565(0.469)(-\sin(0.469x - 2.293))$$

Solve $T''(x) = 0$ to find that the inputs at the inflection points are $x \approx 4.89$ and $x \approx 11.59$. The most rapid increase in the after-tax profit rate was 0.56 percent per year (in 1993), and the most rapid decrease in the rate was 0.55 percent per year (in 2000).

c. *One possible answer:* The smallest after-tax profit rate occurred in 1990. All after-tax profit rates during the 1990s were larger than the after-tax profit rates in the late 1980s.

15. a. $\text{period} = \dfrac{2\pi}{0.01345} \approx 467.2$ milliseconds

b. $p'(s) = 40.5(0.01345)\cos(0.01345s - 1.5708)$ counts per second per millisecond
$p'(s) = 0$ when $s \approx 2.723$, $s \approx 233.576$, $s \approx 467.152$, and $s \approx 700.727$
The pulse speed was at its highest value, 227 counts per second, after approximately 233.6 milliseconds and 700.7 milliseconds. The pulse speed was at its lowest value, 146 counts per second, after approximately 2.7 milliseconds and 467.2 milliseconds.

c. $p''(s) = -40.5(0.01345)(0.01345)\sin(0.01345s - 1.5708)$ counts/second/millisecond^2
$p''(s) = 0$ when $s \approx 116.8$ and $s \approx 350.4$
$p'(116.8) \approx 0.545$ and $p(116.8) \approx 186.5$; $p'(350.4) \approx -0.545$ and $p(350.4) \approx 186.5$
The speed of the pulses emitted from the star was increasing the fastest at approximately 116.791 milliseconds. The speed at that time was approximately 186.5 counts per second, the speed of the pulses emitted from the star was decreasing the fastest after approximately 350.368 milliseconds when the speed was also 186.5 counts per second.

17. a. Using technology, a model is $d(t) = 60.7407 \sin(0.1206t + 0.3134) + 99.9194$ deaths per 100,000 people per week from 1923 through 1925, where t is the number of weeks since January 1, 1923.

b. $d'(t) = 60.7407(0.1206) \cos(0.1206t + 0.3134)$ weekly deaths/100,000 people per week
$d'(t) = 0$ and $d(t)$ is a maximum first beyond the data given when $t \approx 166.8$ weeks after January 1, 1923. Thus the next peak beyond the data given is approximately 166.8 weeks after January 1, 1923.
One possible answer: The peaks in the pneumonia death rate occur approximately every 3 years. Variations in weather might cause the actual peak to differ from the one predicted by the equation.

c. $d''(t) = -60.7407(0.1206)(0.1206) \sin(0.1206t + 0.3134)$ weekly deaths/100,000 people per week per week
$d''(t) = 0$ first beyond the data given when $t \approx 153.70$ weeks after January 1, 1923. The first time after 150 weeks past January 1, 1923 that the weekly pneumonia death rate was increasing the fastest was approximately 153.7 weeks. The weekly death rate at that time was approximately $d(153.7) \approx 100$ deaths per 100,000 people per week.

19. a. $a''(m) = -27.1(0.485)(0.485) \sin(0.485x - 1.707)$ watts per centimeter squared per month per month when $m = 1$ in January, $m = 2$ in February, and so on.

b. Between $x = 60$ and $x = 61$, $a''(m) = 0$ and $a'(m) > 0$ when $m \approx 3.5$ and $a''(m) = 0$ and $a'(m) < 0$ when $m \approx 10.0$.

Radiation is increasing most rapidly in mid-April and decreasing most rapidly at the beginning of October.

c. $a(3.5196) \approx 32.9$ and $a(9.9971) \approx 32.9$

The radiation received at the times in part *b* is 32.9 watts per centimeter squared.

21. a. $R(x) = 0.7c(x) + 0.2p(x)$

$= 1.32727 \sin(0.0186x + 1.1801) + 1.46052 \cos(0.0197x - 3.7526) + 7.90076$

million dollars is the revenue for the Campbell Soup Company from sales of canned soup and powdered drink mix, where x is the day of the year

b. $R'(x) = 1.32727(0.0186)\cos(0.0186x + 1.1801) -$
$1.46052(0.0197)\sin(0.0197x - 3.7526)$
$= 0.0247\cos(0.0186x + 1.1801) - 0.0288\sin(0.0197x - 3.7526)$
million dollars per day, where x is the day of the year

c. For x between 0 and 365, $R'(x) = 0$ and $R(x)$ is maximum when $x \approx 192.95$.
$R(192.95) \approx 8.03$; The highest revenue is approximately $8.03 million on July 12.

d. For x between 0 and 365, $R'(x) = 0$ and $R(x)$ is minimum when $x \approx 66$ and $x \approx 318$. $R(66.27226) \approx 7.6625$, $R(317.6030) \approx 7.6832$

The lowest revenue is approximately $7.66 million on March 7. There is also a local minimum with a revenue of approximately 7.68 million on November 14.

Section 7.5 Accumulation in Cycles

1. $\int_A^x f(t)\,dt$

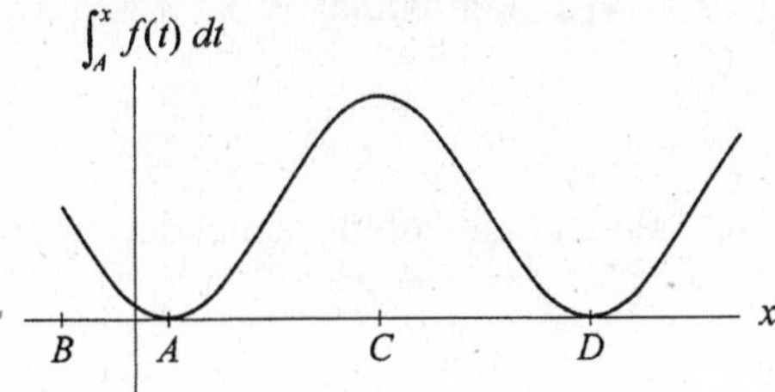

3. $\int_0^a f(x)dx = 1$

5. $\int_{-a}^{a}[f(x)+1]dx = \int_{-a}^{a} f(x)dx + \int_{-a}^{a} dx = 2+2a$

7. $\int(7.3\sin x+12)dx = -7.3\cos x+12x+C$

9. $\int \sin(7.3x+12)dx = \frac{-1}{7.3}\cos(7.3x+12)+C$

11. $\int[4.67\sin(0.024x+3.211)+14.63]dx = \frac{-4.67}{0.024}\cos(0.024x+3.211)+14.63x+C$

$\approx -194.583\cos(0.024x+3.211)+14.63x+C$

13. **a.** Height units are dollars per year, and width units are years. Area units are dollars.

b. The t-intercepts of w between $t = 7$ and $t = 50$ are $t \approx 21.74$ and $t \approx 43.41$. The area between $t = 7$ and $t \approx 21.74$ is \$1.71, the area between $t \approx 21.74$ and $t \approx 43.41$ is \$2.23, and the area between $t \approx 43.41$ and $t = 50$ is \$0.47. The total area is approximately \$4.41.

c.  $\int_7^{50} w(t)dt \approx -0.04$ dollars, which is not the same as the result of part b because w crosses the t-axis. The federal minimum wage, expressed in constant 2000 dollars, decreased by approximately 4 cents between 1957 and 2000.

d.

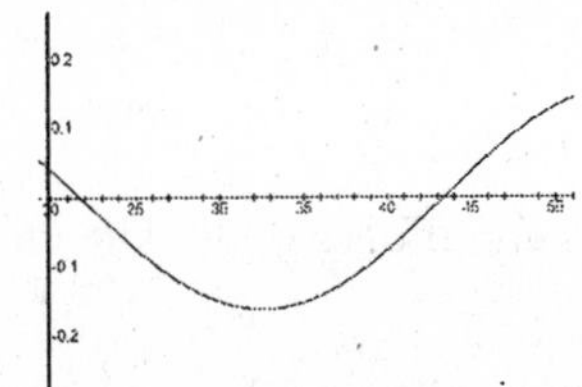

The definite integral is graphed as the area between the curve and the horizontal axis.

e. The federal minimum wage, expressed in constant 2000 dollars, decreased by approximately 2.16 dollars between 1970 and 1995.

15. a. $T(x) = -1.205\cos(0.469x - 0.722) + 6.404$ percent x years after 1988

b. The after-tax profit rate was lowest in 1990 ($x \approx 1.54$) at approximately 5.2% and was highest in 1996 ($x \approx 8.24$) at approximately 7.6%.

c. Approximately 6.4%

d. The values are close, but not identical, because a model (and not the actual data) was used to determine the average value in part *c*.

17. a. $$\int_1^{16} [15.388\sin(0.787x - 1.667) + 16.346]dx = \left[\frac{-15.388}{0.787}\cos(0.787x - 1.667) + 16.346x\right]_1^{16}$$

$$\approx 262.915 - 3.888 \approx 259 \text{ thousand lizards}$$

b. $\sum_{x=1}^{16} L(x) \approx 261$ thousand lizards. This value is 2 thousand more than the answer to part *a*.

c. *One possible answer:* Using a calculator's definite integral function, the definite integral is quickly evaluated.

19. $\int_0^{40} r(t)dt \approx -4.2$

From 1900 to 1940, the ratio in the number of males per 100 females decreased by approximately
4.2 males per 100 females.

$\int_{50}^{100} r(t)dt \approx -2.33$

From 1950 to 2000, the ratio in the number of males per 100 females decreased by approximately
2.33 males per 100 females.

$\int_0^{100} r(t)dt \approx -8.8$

From 1900 to 2000, the ratio in the number of males per 100 females decreased by approximately
8.8 males per 100 females.

21. a. The height units are counts per second. The width units are milliseconds. The units for the area are (counts per second)(millisecond).

b. $p(s) = 0.0405\sin(0.01345s - 1.5708) + 0.1865$ counts per millisecond after s milliseconds

c. $$\int_0^{467.151324} [0.0405\sin(0.0135s - 1.5708) + 0.1865]ds$$

$$= \left(\frac{-0.0405}{0.0135}\cos(0.0135s - 1.5708) + 0.1865s\right)\Bigg|_0^{467.151324} \approx 87$$

Approximately 87 pulses are emitted by the star over one period (approximately 467 milliseconds or approximately 0.5 second).

Chapter 7 Concept Review

1. a. Both vertical shift and amplitude are based on the highest and lowest output values. In the table, the highest value is 1500, and the lowest value is 75.

$$\text{Vertical shift} \approx \frac{1500+75}{2} = 787.5$$

$$\text{Amplitude} \approx \frac{1500-75}{2} = 712.5$$

Period $\approx$ 12 months

To determine the horizontal shift, we find the first output value closest to the vertical shift. In this case, the March value of 800 lawn mowers fits this description. Thus we conclude that

Horizontal shift ≈ 3 months to the right.

b. Using technology, a sine model is $l(x) = 713.2507 \sin(0.5250x - 1.5569) + 777.8826$ lawn mowers ordered x months since December of the previous year. The amplitude is very close, whereas the vertical shift of the model is approximately 10 lawn mowers less than that found using the data. The period of the model is approximately 12 months. The horizontal shift and period are very close to those found using the data.

c. $l(16) \approx 1157$ lawnmowers. For this estimate to be valid, the cyclic pattern shown by the data should continue for the following year.

d. Using technology, we find that the model appears to have greatest number of orders when $x = 6$ (June).

e. Find the value of x for which $l''(x) = 0$ and $l'(x) > 0$. Using technology, $x \approx 3$ (March).

2. a. Average rate of change $\approx \dfrac{56-103}{30} \approx -1.6$ units per year

b. Estimate the slope to be approximately units per year at 1860.

3. a. Average rate of change $= \dfrac{S(72)-S(42)}{30} \approx -1.32$ units per year

b. $S'(t) = 11.2(0.0654)\cos(0.0654t + 5.9690) + 13.8(0.1309)\cos(0.1309t + 0.9599)$
$+ 5.3(0.3272)\cos(0.3272t + 0.9076)$ per year t years after 1818

$S'(42) \approx 0.37$ unit per year

According to the model, the Sauerbeck index was increasing by approximately 0.37 unit per year in 1860.

c. $$\frac{1}{72-42}\int_{42}^{72} S(t)dt = \frac{1}{30}\left(88.6t - \frac{11.2}{0.0654}\cos(0.0654t + 5.9690) - \frac{13.8}{0.1309}\cos(0.1309t + 0.09599) - \frac{5.3}{0.3272}\cos(0.3272t + 0.9076\right)\Bigg|_{72}^{42} \approx 90.8 \text{ unit per year}$$

4. a. $r'(t) = -0.1439(0.0197)\cos(0.0197t - 3.7526)$ million pints per day per day t years after January 1, 1992.

Using technology, $r'(t) = 0$ and $r''(t) < 0$ when $t \approx 111$ days, so the rate of change was greatest on March 20 (1992 was a leap year).

b. $R(t) = \int -0.1439\sin(0.0197t - 3.7526)dt = \dfrac{0.1439}{0.0197}\cos(0.0197t - 3.7526) + C$

Because $R(366) = 14.4$, we solve for C to get $C \approx 20.383$.

$R(t) = 7.3046\cos(0.0197t - 3.7526) + 20.383$ million pints t days after January 1, 1992.

c. The answer to part *a* is the input corresponding to the inflection point for the model in part *b*. This is where the slope of the graph of the equation in part *b* is the steepest.

5. $\int_{32}^{61} -0.1439\sin(0.0197t - 3.7526)dt = \dfrac{0.1439}{0.0197}\cos(0.0197t - 3.7526)\Big|_{32}^{61} \approx 1.24$ million pints

The number of pints of powdered drink mix sold daily by Campbell's increased by 1.24 million pints from the beginning of February, 1992 through the end of February, 1992.

Chapter 8
Dynamics of Change: Differential Equations and Proportionality

Section 8.1 Differential Equations and Slope Fields

1. $c = kg$

3. $\frac{dp}{da} = ka$

5. a.

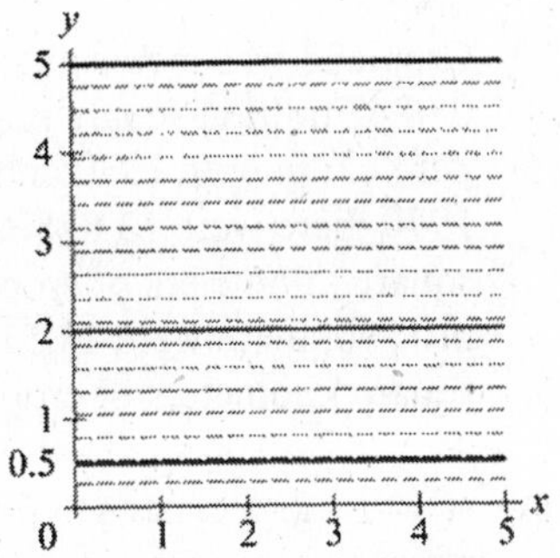

One possible answer: The displayed particular solutions go through the points (0, 0.5), (2, 2) and (4, 5).

b. All particular solutions are horizontal lines, each passing through the chosen initial condition.

c. $y = C$, where C is a constant

7. a.

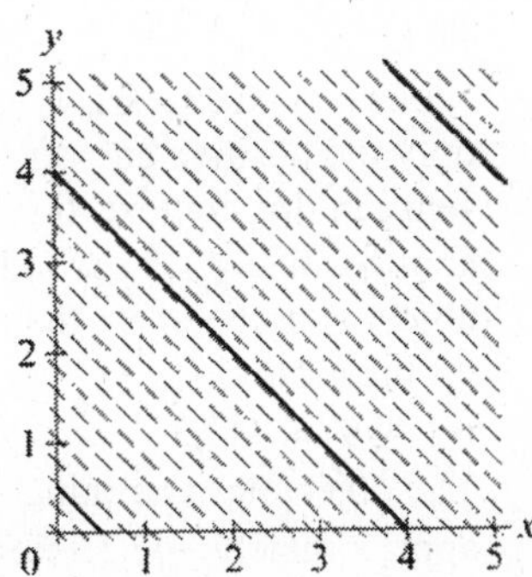

One possible answer: The displayed particular solutions go through the points (0, 0.5), (2, 2) and (4, 5).

b. All particular solutions are parallel lines with slope –1 and differing vertical shifts.

c. $y = -x + C$, where C is a constant

9. a.

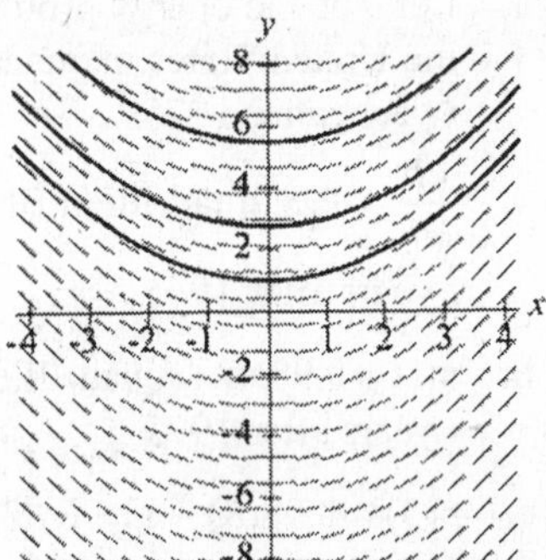

One possible answer: The displayed particular solutions go through the points (0, 1), (1, 3) and (2, 6.5).

b. All particular solutions are concave-up parabolas with minimum points on the vertical axis.

c. $y = 0.25x^2 + C$, where C is a constant

11. a.

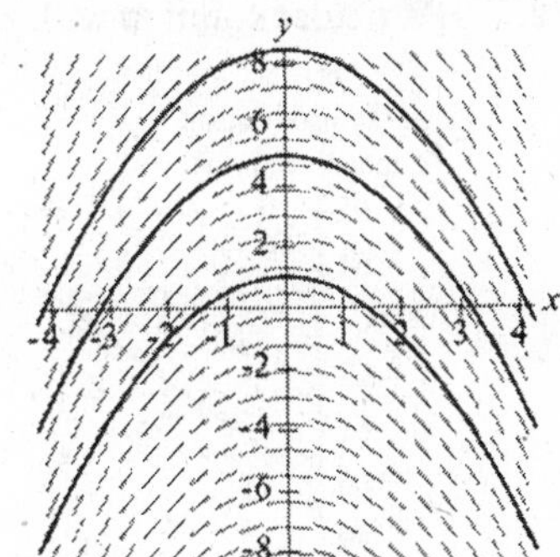

One possible answer: The displayed particular solutions go through the points (0, 1), (2, 3) and (2, 6.5).

b. All particular solutions are concave-down parabolas with maximum points on the vertical axis.

c. $y = -0.5x^2 + C$, where C is a constant

13. The particular solutions with initial condition (0, 0) all pass through the point (0,0). Each particular solution has a constant term zero, but the nonconstant terms all differ.

15. a. Let p be the energy production in the United States in quadrillion Btu t years after 1975.

$\frac{dp}{dt} = 0.98$ quadrillion Btu/year t years after 1975.

b. $p(t) = 0.98t + C$ quadrillion Btu t years after 1975

c. $p(5) = 0.98(5) + C = 64.8$, so $C = 59.9$

$p(t) = 0.98t + 59.9$ quadrillion Btu t years after 1975

d. $p(0) = 59.9$ and $\frac{dp}{dt} = 0.98$ when $t = 0$.

In 1975 the production was 59.9 quadrillion Btu and was increasing at a rate of 0.98 quadrillion Btu per year.

e. Particular solution with $p(0)$=59.9

17. a. Let c be the amount of arable and permanent cropland in millions of square kilometers t years after 1970.

$\frac{dc}{dt} = 0.0342$ million square kilometers per year t years after 1970

b. $c(t) = 0.0342t + C$ million square kilometers t years after 1970

c. $c(10) = 0.0342(10) + C = 14.17$, so $C = 13.828$

$c(t) = 0.0342t + 13.828$ million square kilometers t years after 1970

d. $c(0) = 13.828$, $c(20) = 14.512$ and $\frac{dc}{dt} = 0.0342$ when $t = 0$ and $t = 20$.

Cropland was increasing at a rate of 0.0342 million square kilometers per year in both 1970 and 1990. In 1970 there were 13.828 million square kilometers of cropland, and in 1990, there were 14.512 million square kilometers of cropland.

19. a. $v(t) = -32t$ feet per second t seconds after the object is dropped

b. Let s be distance.

$\frac{ds}{dt} = -32t$ feet per second t seconds after the object is dropped

c. $s(t) = -16t^2 + C$ feet t seconds after the object is dropped

d. $s(0) = -16(0^2) + C = 35$, so $C = 35$

$s(t) = -16t^2 + 35$

When $s(t) = 0$, $t \approx \pm 1.479$, where only the positive answer makes sense in this context. It takes approximately 1.5 seconds after the object is dropped for the object to hit the ground.

$v(1.479) \approx -47.3$

The object has a terminal velocity of approximately –47.3 feet per second. (The negative sign on the velocity indicates downward motion.)

21. a. $\dfrac{df}{dx} = kx$

b. $f(x) = \dfrac{k}{2}x^2 + C$

c. Taking the derivative of f, we get

$$\frac{d}{dx}\left(\frac{k}{2}x^2 + C\right) = \frac{k}{2}(2x) + 0 = kx$$

Thus we have the identity $kx = kx$, and our solution is verified.

23. a. $\dfrac{dh}{dt} = \dfrac{k}{t}$ feet per year after t years

b. $h(t) = k\ln|t| + C$ feet in t years

$h(2) = k\ln 2 + C = 4$ and

$h(7) = k\ln 7 + C = 30$

Solving this system of equations, we get $k \approx 20.75$ and $C \approx -10.39$.

$h(t) = 20.75\ln t - 10.39$ feet in t years

c. $h(15) \approx 45.8$ feet

Over time, the tree will continue to grow, but the rate of increase will be smaller and smaller.

25. a.

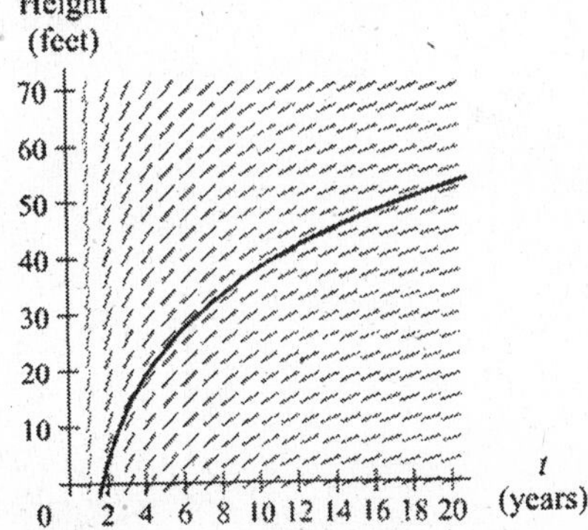

b. The output value on the graph corresponding to an input of $t = 15$ is $h(15) \approx 46$ feet.

27. a.

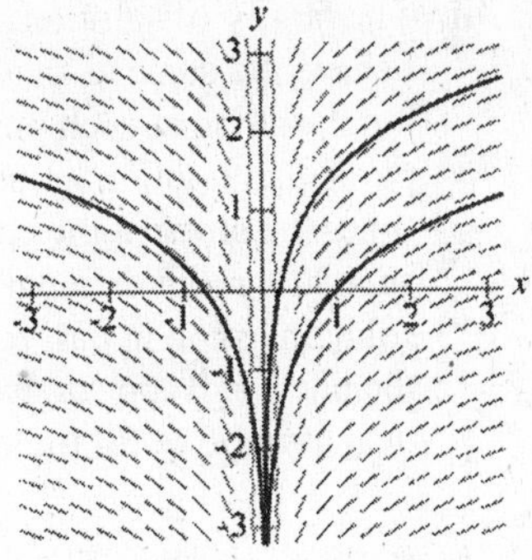

i. *One possible answer:* The particular solutions shown go through (1, 1.5), (–2, 1), and (1, 0).

ii. When $x > 0$, the graph of a particular solution rises as x gets larger. When $x < 0$, the solution graph rises as x gets smaller. The particular solution graphs are concave down.

iii. The family of solutions appears to increase rapidly as x moves away from the origin (in both directions), and then the increase slows down. The line $x = 0$ (lying on the y-axis) appears to be a vertical asymptote for the family.

b.

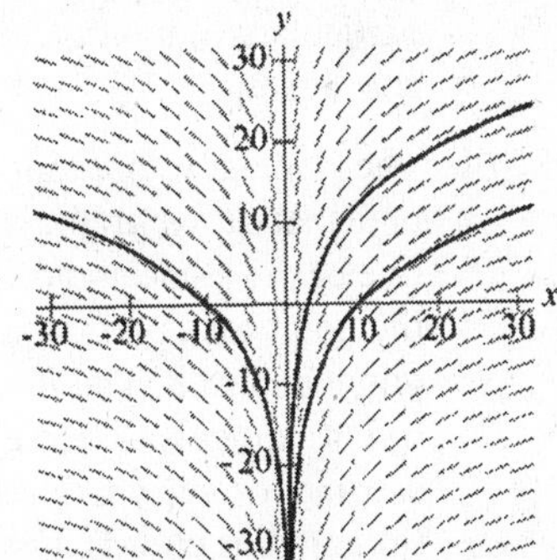

i. *One possible answer:* The particular solutions shown go through (10, 0), (–10, 0), and (5, 5).

ii. When $x > 0$, the graph of a particular solution rises as x gets larger. When $x < 0$, the solution graph rises as x gets smaller. The particular solution graphs are concave down.

iii. The family of solutions appears to behave the same as that in part *a*, but the slope at each point on a particular solution graph is 10 times the slope at the corresponding point on a particular solution graph in part *a*. Again, the line $x = 0$ appears to be a vertical asymptote for the family.

c.

i. *One possible answer:* The particular solutions shown go through (1, 1.5), (−2, 1), and (1, 0).

ii. When $x > 0$, the graph of a particular solution falls as x gets larger. When $x < 0$, the solution graph falls as x gets smaller. The particular solution graphs are concave up.

iii. The slope at each point on a particular solution graph is the negative of the slope at a corresponding point on a particular solution graph in part *a*. The family of solutions appears to decrease rapidly as x moves away from the origin (in both directions), and then the decrease levels off. The line $x = 0$ appears to be a vertical asymptote for the family.

d.

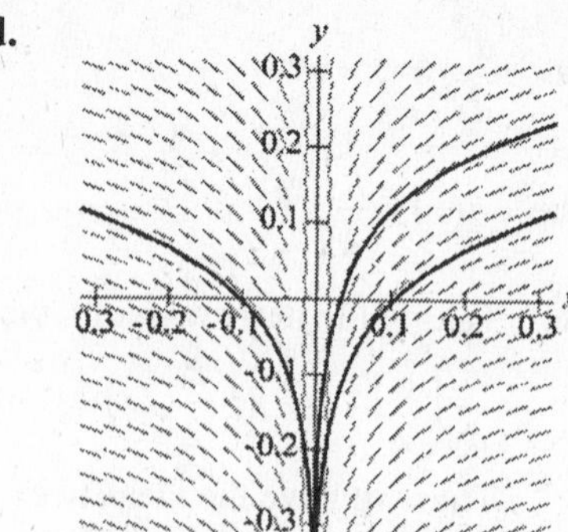

i. *One possible answer:* The particular solutions shown go through (0.1, 0), (−0.1, 0), and (0.05, 0.05).

ii. When $x > 0$, the graph of a particular solution rises as x gets larger. When $x < 0$, the solution graph rises as x gets smaller. The particular solution graphs are concave down.

iii. The family of solutions appears to behave the same as that in part *a*, but the slope at each point on a particular solution graph is $\frac{1}{10}$ times the slope at the corresponding point on a particular solution graph in part *a*. Again, the line $x = 0$ appears to be a vertical asymptote for the family.

29.

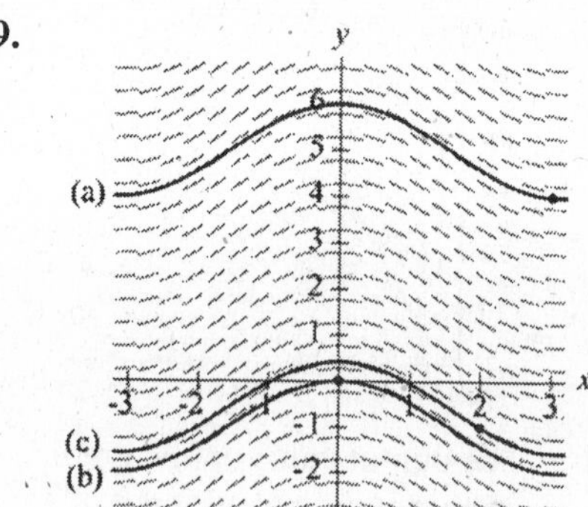

31.

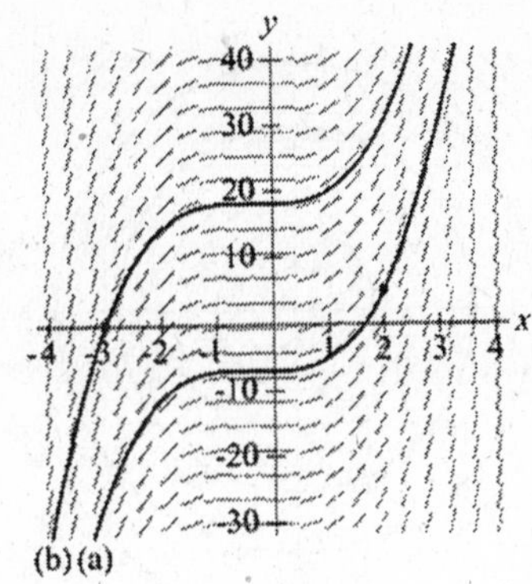

Section 8.2 Separable Differential Equations

1. $\frac{dT}{dt} = \frac{k}{T}$

This is a separable differential equation.

$$TdT = kdt$$

$$\int TdT = \int kdt$$

$$\tfrac{1}{2}T^2 + c_1 = kt + c_2$$

$$T^2 = 2kt + C$$

$$T = \pm\sqrt{2kt + C}$$

Because thickness can't be negative, $T(t) = \sqrt{2kt + C}$.

3. $\frac{dA}{dt} = kA$

This a separable differential equation.

$$\frac{1}{A}dA = kdt$$

$$\int \frac{1}{A}dA = \int kdt$$

$$\ln A + c_1 = kt + c_2$$

$$\ln A = kt + C$$

$$A = e^{kt+C} \quad \left(\text{note that } e^{kt+C} = e^{kt} \cdot e^{C}\right)$$

$$A = ae^{kt} \quad \left(\text{where } a = e^{C}\right)$$

(Note that A is positive, so we omitted the absolute value signs from the natural logarithms,)

$$A(t) = ae^{kt}$$

5. $\frac{dx}{dt} = kx(N - x)$

This differential equation of the form $\frac{dP}{dt} = kP(C - P)$ has a logistic function of the form $P(t) = \frac{C}{1 + Ae^{-Ckt}}$ as its general solution.

Thus $x(t) = \frac{N}{1 + Ae^{-Nkt}}$

7. Flow-in rate =

$\frac{k}{\sqrt{D}}$ where k is a constant

Flow-out rate =

hD where h is a constant

$$\frac{dD}{dt} = \frac{k}{\sqrt{D}} - hD$$

9.

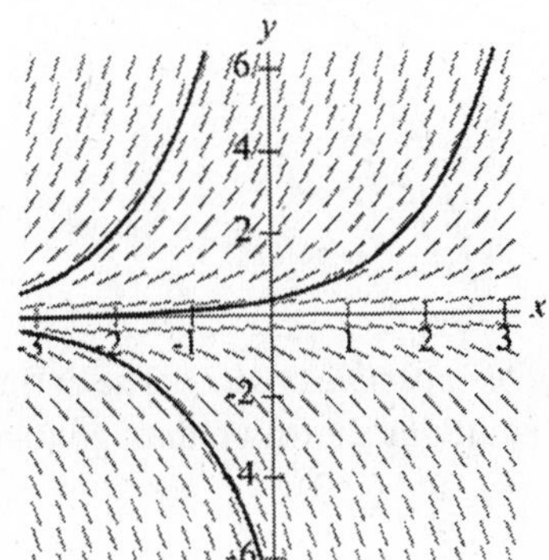

One possible answer: The particular solutions shown go through (–2, 2), (–2, –1), and (1, 1).

11.

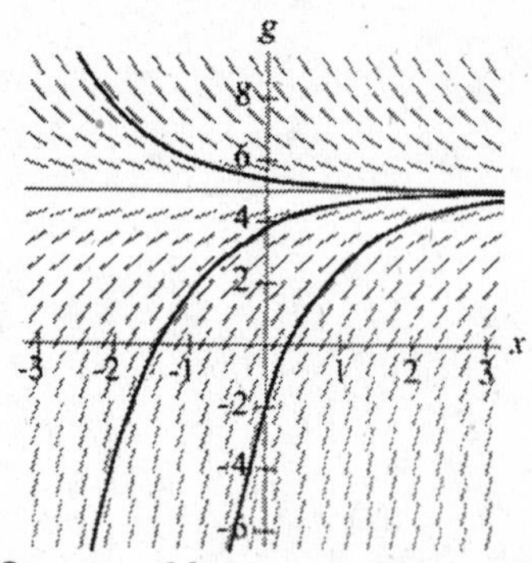

One possible answer: The particular solutions shown go through (0, –2), (–2, –4), and (–1, 6).

13.

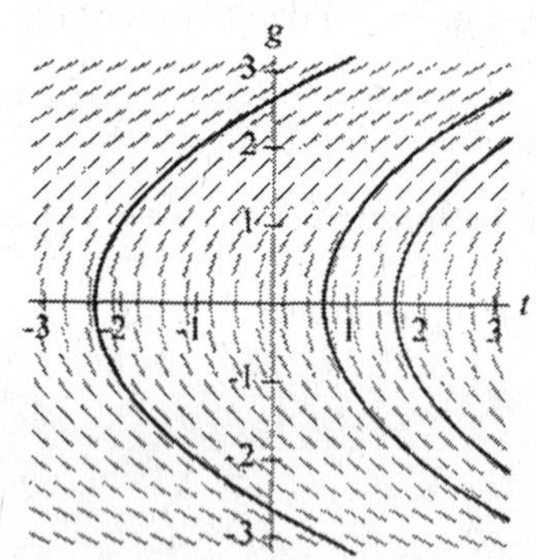

One possible answer: The particular solutions shown go through (–2, 1), (2, 1), and (1, –1).

15.

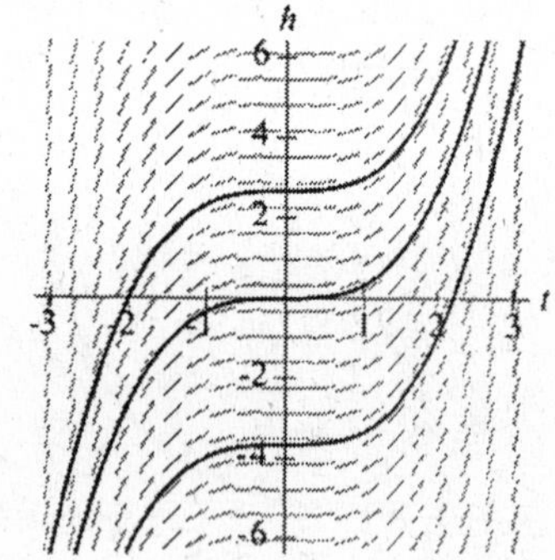

One possible answer: The particular solutions shown go through (0, 0), (–1, –4), and (1, 3).

17.

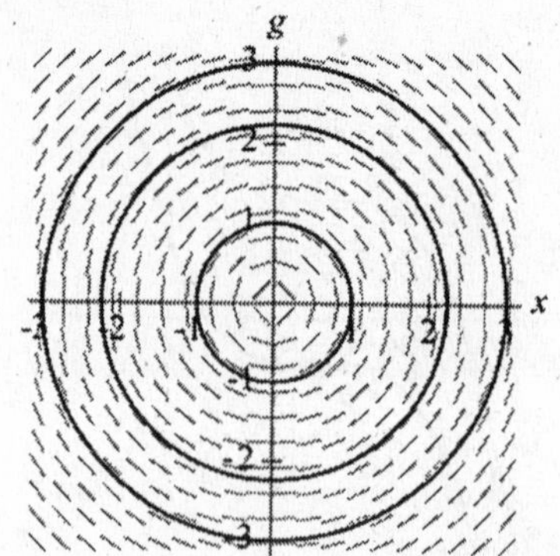

One possible answer: The particular solutions shown go through (0, 3), (1, 0), and (2, 1).

19. Solve by separation of variables.

$$\frac{1}{y}dy = kdx$$

$$\int \frac{1}{y}dy = \int kdx$$

$$\ln|y| + c_1 = kx + c_2$$

$$\ln|y| = kx + C$$

$$y = \pm e^{kx+C}$$

$$y(x) = \pm ae^{kx}$$

21. Solve by antidifferentiation.

$$\int dy = \int \frac{k}{x}dx$$

$$y(x) = k\ln|x| + C$$

22. Solve by separation of variables.

$$ydy = kxdx$$
$$\int ydy = \int kxdx$$
$$\tfrac{1}{2}y^2 + c_1 = \tfrac{k}{2}x^2 + c_2$$
$$y^2 = kx^2 + C$$
$$y(x) = \pm\sqrt{kx^2 + C}$$

23. Solve by separation of variables.

$$\frac{1}{y}dy = \frac{k}{x}dx$$
$$\int \frac{1}{y}dy = \int \frac{k}{x}dx$$
$$\ln|y| + c_1 = k\ln|x| + c_2$$
$$\ln|y| = k\ln|x| + C$$
$$|y| = e^C e^{\ln|x|^k}$$
$$|y| = a|x|^k$$
$$y(x) = \pm ax^k$$

(Note that $k\ln|x| + C$ can be rewritten as $\ln|x|^k + C$. Now change the equation from logarithmic form to exponential form. $|y| = e^{k\ln|x|+C} = e^{k\ln|x|} \cdot e^C$. Replace e^C with a and remove the absolute value signs.)

25. a. $\dfrac{dq}{dt} = kq$ milligrams per hour

b. Solve by separation of variables.

$$\tfrac{1}{q}dq = kdt$$
$$\int \tfrac{1}{q}dq = \int kdt$$

$$\ln|q| + c_1 = kt + c_2$$
$$\ln|q| = kt + C$$
$$q = e^{kt+C}$$
$$q(t) = ae^{kt}$$

When $t = 0$, $q = 200$, and when $t = 2$, $q = 100$. Thus, we have

$q(0) = a = 200$ and $q(2) = ae^{2k} = 100$

$$200e^{2k} = 100$$
$$e^{2k} = \tfrac{1}{2}$$
$$2k = \ln\tfrac{1}{2}$$
$$k = \tfrac{1}{2}\ln\tfrac{1}{2}$$

Thus we have $a = 200$ and $k = \frac{1}{2}\ln\frac{1}{2} \approx -0.346574$.

$q(t) = 200e^{-0.346574t}$ milligrams
after t hours

c. After 4 hours, $q(4) \approx 50$ milligrams will remain. After 8 hours, $q(8) \approx 12.5$ milligrams will remain.

27. a. $\dfrac{da}{dt} = ka$ units per day

b. Solve by separation of variables.

$$\frac{1}{a}da = kdt$$

$$\int\frac{1}{a}da = \int kdt$$

$$\ln|a| + c_1 = kt + c_2$$

$$\ln|a| = kt + C$$

$$a = e^{kt+C}$$

$$a(t) = ce^{kt}$$

$a(0) = c =$ the initial amount.
When $t = 3.824$,
$a =$ half of the initial amount $= \frac{c}{2}$. Thus, we have

$$\frac{c}{2} = ce^{3.824k}$$

$$\frac{1}{2} = e^{3.824k}$$

$$\ln\frac{1}{2} = 3.824k$$

$$k = \frac{1}{3.824}\ln\frac{1}{2}$$

$$\approx -0.181262$$

$a(t) = ce^{-0.181262t}$ units after t days

c. Here, $c = 1$ gram.
After 12 hours = 0.5 days,
$a(0.5) \approx 0.91$ gram will remain. After 4 days, $a(4) \approx 0.48$ gram will remain. After 9 days, $a(9) \approx 0.20$ gram will remain. After 30 days, $a(30) \approx 0.004$ grams will remain.

29. a. Let N be number of countries that issued stamps.

$\dfrac{dN}{dt} = 0.0049N(37 - N)$ countries per year

b. This differential equation has a logistic function as its solution.

$$N(t) = \frac{37}{1 + Ae^{-0.0049(37)t}}$$

$$= \frac{37}{1 + Ae^{-0.1813t}} \text{ countries}$$

t years after 1800

c. Because $N(55) = 16$, we have the equation $N(55)=\dfrac{37}{1+Ae^{-0.1813(55)}}=16$. Solving this equation for A, we get

$$\frac{37}{1+Ae^{-0.1813(55)}}=16$$
$$37=16+16Ae^{-9.9715}$$
$$A=\frac{21}{16e^{-9.9715}}$$
$$\approx 28{,}097.439$$

$N(t)=\dfrac{37}{1+28{,}097.439e^{-0.1813t}}$ countries t years after 1800

d. $N(40)\approx 2$ and $N(60)\approx 24$; There were 2 countries in 1840 and 24 countries in 1860.

e. The upper asymptote is $N(t)=37$, and the lower asymptote is $N(t) = 0$.

f.

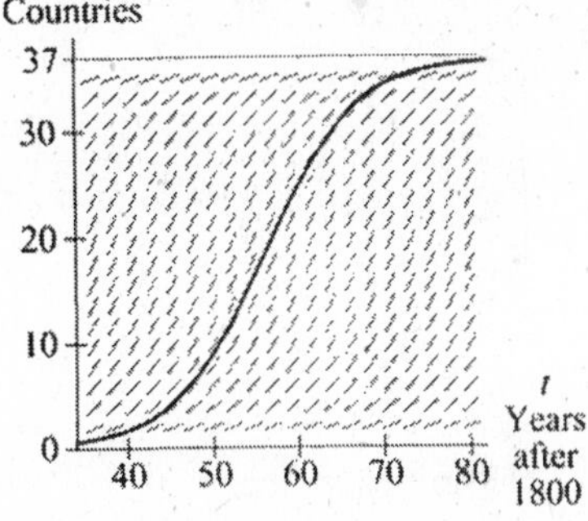

31. a. $\dfrac{df}{dx}=kf(L-f)$

b. This differential equation has a logistic function as its solution: $f(x)=\dfrac{L}{1+Ae^{-Lkx}}$

c. Taking the derivative in part b, we get

$$\frac{df}{dx}=-L(1+Ae^{-Lkx})^{-2}Ae^{-Lkx}(-Lk)$$
$$=\frac{-L}{(1+Ae^{-Lkx})^2}(-LkAe^{-Lkx})=\frac{L^2kAe^{-Lkx}}{(1+Ae^{-Lkx})^2}$$

Substituting $f(x)=\dfrac{L}{1+Ae^{-Lkx}}$ into $kf(L-f)$ and simplifying gives

$$kf(L-f) = k\left(\frac{L}{1+Ae^{-Lkx}}\right)\left(L-\frac{L}{1+Ae^{-Lkx}}\right)$$
$$= k\left(\frac{L}{1+Ae^{-Lkx}}\right)\left(\frac{L(1+Ae^{-Lkx})}{1+Ae^{-Lkx}}-\frac{L}{1+Ae^{-Lkx}}\right)$$
$$= k\left(\frac{L}{1+Ae^{-Lkx}}\right)\left(\frac{L+LAe^{-Lkx}-L}{1+Ae^{-Lkx}}\right) = \frac{kL^2Ae^{-Lkx}}{\left(1+Ae^{-Lkx}\right)^2}$$

Thus we have the identity $\frac{kL^2Ae^{-Lkx}}{\left(1+Ae^{-Lkx}\right)^2} = \frac{kL^2Ae^{-Lkx}}{\left(1+Ae^{-Lkx}\right)^2}$, and our solution is verified.

Section 8.3 Numerically Estimating by Using Differential Equations: Euler's Method

1. a. $\frac{dy}{dx} = \frac{1}{2}$, initial condition (0, 0), step size $\frac{4-0}{2 \text{ steps}} = 2$ units per step

x	**Estimate of $y(x)$**	**Slope at x**
0	0	0.5
2	$0+2(0.5)=1$	0.5
4	$1+2(0.5)=2$	

Thus we estimate $y(4) \approx 2$.

b. $\frac{dy}{dx} = 2x$, initial condition (1, 4), step size $\frac{7-1}{2 \text{ steps}} = 3$ units per step

x	**Estimate of $y(x)$**	**Slope at x**
1	4	$2(1)=2$
4	$4+3(2)=10$	$2(4)=8$
7	$10+3(8)=34$	

Thus we estimate $y(7) \approx 34$.

3. a. $\frac{dy}{dx} = \frac{5}{y}$, initial condition (1, 1), step size is $\frac{5-1}{2} = 2$ units per step

x	**Estimate of $y(x)$**	**Slope at x**
1	1	$\frac{5}{1}=5$
3	$1+2(5)=11$	$\frac{5}{11}\approx 0.4545$
5	$11+2\left(\frac{5}{11}\right)\approx 11.91$	

We estimate that $y(5) \approx 11.91$.

b. $\frac{dy}{dx} = \frac{5}{x}$, initial condition (2, 2), step size is $\frac{8-2}{2} = 3$ units per step

x	Estimate of $y(x)$	Slope at x
2	2	$\frac{5}{2} = 2.5$
5	$2 + 3\left(\frac{5}{2}\right) = 9.5$	$\frac{5}{5} = 1$
8	$9.5 + 3(1) = 12.5$	

We estimate that $y(8) \approx 12.5$.

5. a. $\frac{dw}{dt} = \frac{33.67885}{t}$ pounds per month after t months

b. Initial condition (1, 6), step size 0.25 month

t (months)	Estimate of $w(t)$ (pounds)	Slope at t (pounds per month)	t (months)	Estimate of $w(t)$ (pounds)	Slope at t (pounds per month)
1	6.000	33.679	3.25	48.768	10.363
1.25	14.420	26.943	3.5	51.359	9.623
1.5	21.155	22.453	3.75	53.764	8.981
1.75	26.769	19.245	4	56.010	8.420
2	31.580	16.839	4.25	58.115	7.924
2.25	35.790	14.968	4.5	60.100	7.484
2.5	39.532	13.472	4.75	61.967	7.090
2.75	42.900	12.247	5	63.739	6.736
3	**45.961**	11.226	5.5	67.027	6.123
3.25	48.768	10.363	5.75	68.558	5.857
3.5	51.359	9.623	**6**	**70.022**	

The Euler estimates using a step size of 0.25 month are $w(3) \approx 46.0$ lbs; $w(6) \approx 70.0$ lbs

c. Initial condition (1, 6), step size 1 month $w(3) \approx 56.52$ pounds; $w(6) \approx 82.90$ pounds

t (months)	Estimate of $w(t)$ (pounds)	Slope at t (pounds per month)	t (months)	Estimate of $w(t)$ (pounds)	Slope at t (pounds per month)
1	6	33.679	4	67.745	8.420
2	39.679	16.839	5	76.164	6.736
3	**56.518**	11.226	**6**	**82.90**	

The Euler estimates using a step size of 1 month are $w(3) \approx 56.5$ lbs; $w(6) \approx 82.9$ lbs

d. The answer to part *b* should be more accurate because it uses a smaller step size.

7. a. $\frac{dp}{dt} = 3.935t^{3.55}e^{-1.35135t}$ thousand barrels per year *t* years after production begins, where $p(t)$ is the total amount of oil produced after *t* years

b. Initial condition: (0, 0); Step size: 0.5 year

t	Estimate of $p(t)$	$p'(t)$	t	Estimate of $p(t)$	$p'(t)$
0	0	0	3	4.9680	3.3734
0.5	0	0.1709	3.5	6.6547	2.9668
1	0.0855	1.0187	4	8.1381	2.4251
1.5	0.5948	2.1865	4.5	9.3507	1.8744
2	1.6881	3.0891	**5**	**10.2879**	
2.5	3.2326	3.4707			

After 5 years of production, the well has produced approximately 10.3 thousand barrels.

c.

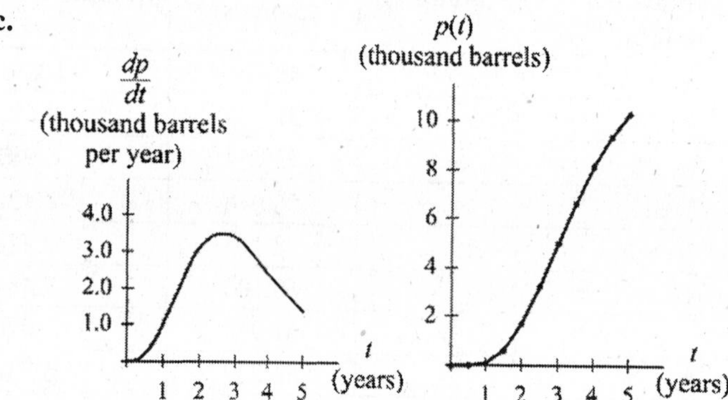

The graph of the differential equation is the slope graph for the graph of the Euler estimates. Similarly, the graph of the Euler estimates is an approximation to the accumulation graph of the differential equation graph.

9. a. $\frac{dT}{dt} = k(T - A)$ °F per minute after *t* minutes.

b. Solve the equation $k(98 - 70) = -1.8$ to get $k \approx -0.064$.

c.

t	Estimate of $p(t)$	$p'(t)$	t	Estimate of $p(t)$	$p'(t)$
0	98	−1.8	8	86.4552	−1.05784
1	96.2	−1.68429	9	85.3974	−0.989833
2	94.5157	−1.57601	10	84.4076	−0.926201
3	92.9397	−1.47470	11	83.4814	−0.866659

4	91.465	−1.37989	12	82.6147	−0.810945
5	90.0851	−1.29119	13	81.8038	−0.758813
6	88.7939	−1.20818	14	81.0449	−0.710032
7	87.5857	−1.13051	**15**	**80.3349**	

After 15 minutes, the temperature of the object is approximately 80.3°F.

11. Euler's method uses tangent-line approximations. Tangent lines generally lie close to a curve near the point of tangency and deviate more and more as you move farther and farther away from that point. Thus, smaller steps generally result in better approximations. Compare the graphs in Activity 4*b*.

13. a,b. With monthly intervals, there are 60 intervals with step size 1/12. The Euler estimate for $p(5)$ is approximately 10.594 thousand barrels.

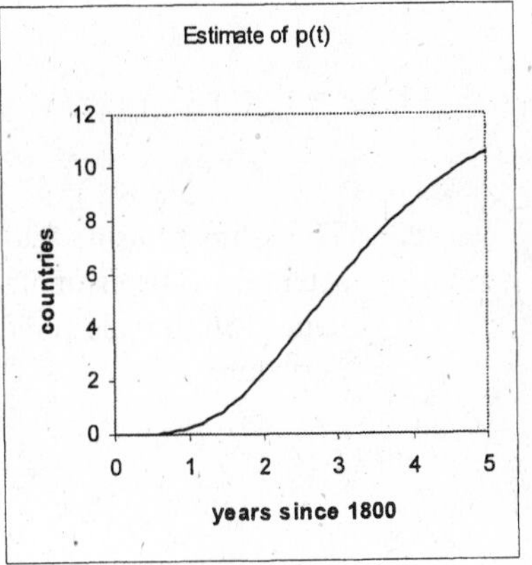

With weekly intervals, there are 260 intervals with step size 1/52. The Euler estimate for $p(5)$ is approximately 10.639 thousand barrels.

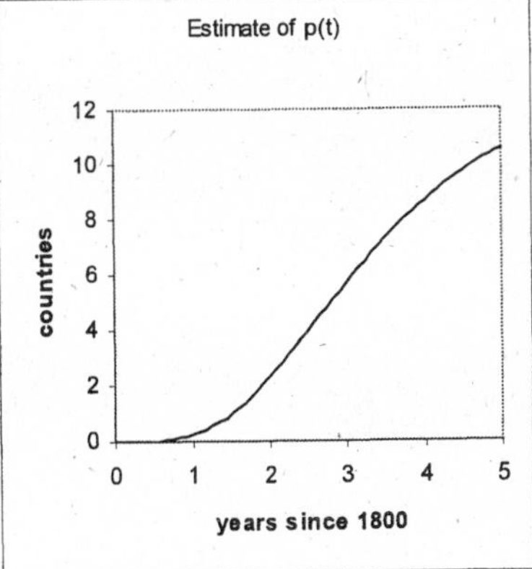

With daily intervals, there are 1825 intervals with step size 1/365. The Euler estimate for $p(5)$ is approximately 10.650 thousand barrels.

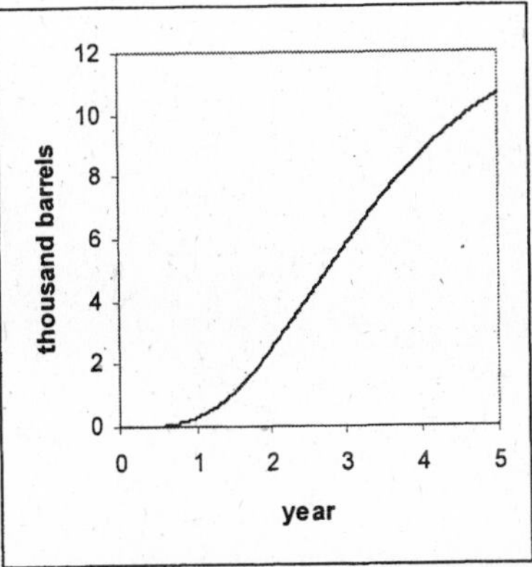

c. *One possible answer:* The estimates using smaller steps are close to the first estimate. Using smaller steps does not greatly change the estimate.

15. a,b. There are 790 one-second intervals with step size 1/60. The Euler estimate for $T(15)$ is approximately 80.6696349 or 80.7 °F.

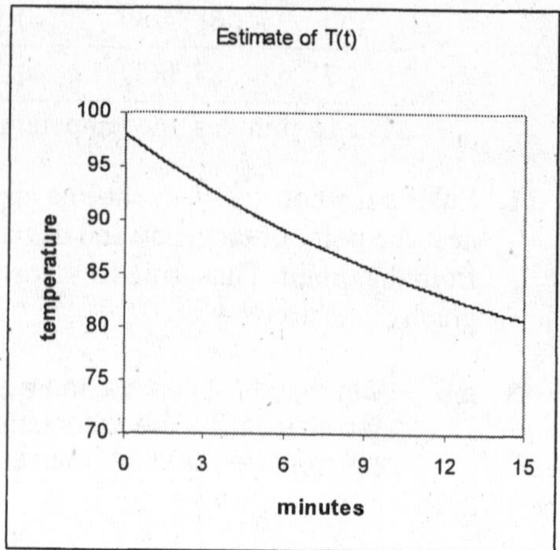

c. The estimate using 790 intervals instead is 0.4 degrees higher than the estimate using 15 intervals. The estimate using 790 intervals is probably more accurate than the estimate using 15 intervals since the function has no really steep areas of descent and no change of curvature.

Section 8.4 Second-Order Differential Equations

1. $\dfrac{d^2s}{dt^2}=\dfrac{k}{s^2}$

3. $\dfrac{d^2P}{dy^2}=k$; Taking the antiderivative, we get $\dfrac{dP}{dy}=ky+C$.

Taking the antiderivative of $\dfrac{dP}{dy}$, we get $P(y)=\dfrac{k}{2}y^2+Cy+D$.

5. a. $\dfrac{d^2R}{dt^2}=6.14$ jobs per month per month in the tth month of the year

b. Taking the antiderivative of $\dfrac{d^2R}{dt^2}$, we get $\dfrac{dR}{dt}=6.14t+C$.

When $t=1$, $\dfrac{dR}{dt}=-0.87$. Solving for C, we get $-0.87=6.14(1)+C$

$C=-7.01$

$\dfrac{dR}{dt}=6.14t-7.01$ jobs per month in the tth month of the year

Taking the antiderivative of $\dfrac{dR}{dt}$, we get $R(t)=3.07t^2-7.01t+C$.

When $t=2$, $R=14$. Solving for C, we get $14=3.07(2^2)-7.01(2)+C$

$C=15.74$

The particular solution is $R(t)=3.07t^2-7.01t+15.74$ jobs in the tth month of the year

c. $R(8)\approx 156$ and $R(11)\approx 310$
We estimate the number of jobs in August to be approximately 156 and the number in November to be 310.

7. a. $\dfrac{d^2A}{dt^2}=-2009$ cases per year per year, where t is the number of years since 1988

b. Taking the antiderivative of $\dfrac{d^2A}{dt^2}$, we get $\dfrac{dA}{dt}=-2099t+C$. When $t=0$, $\dfrac{dA}{dt}=5988.7$, so $C=5988.7$. The particular solution is $\dfrac{dA}{dt}=-2099t+5988.7$ cases per year, where t is the number of years since 1988.

Taking the antiderivative of $\dfrac{dA}{dt}$, we get $A(t)=-1049.5t^2+5988.7t+C$. When $t=0$, $A=33{,}590$, so $C=33{,}590$. The particular solution is
$A(t)=-1049.5t^2+5988.7t+33{,}590$ cases, where t is the number of years since 1988

c. When $t = 3$, $\frac{dA}{dt} = -308.3$ and $A(3) \approx 42{,}111$.

We estimate that in 1991 there were 42,111 AIDS cases and the number of cases was decreasing at rate of 308.3 cases per year.

9. a. $\frac{d^2 f}{dx^2} = k$

b. Taking the antiderivative, we have $\frac{df}{dx} = kx + C$.

Taking the antiderivative of the result, we get $f(x) = \frac{k}{2}x^2 + Cx + D$.

c. Taking the derivative in part *b*, we get $\frac{d}{dx}\left(\frac{k}{2}x^2 + Cx + D\right) = \frac{k}{2}(2x) + C + 0 = kx + C$

Taking the derivative of this result, we get $\frac{d}{dx}(kx + C) = k$ so we have the identity $k = k$, and our solution is verified.

11. a. $\frac{d^2 E}{dt^2} = -0.212531E$ mm per day per month per month, where $E(t)$ is the amount of radiation in mm per day and t is measured in months

b. The general solution to the equation is of the form $E(t) = a\sin(\sqrt{k}t + c)$ with $\sqrt{k} = \sqrt{0.212531} \approx 0.461011$.

In June, the amount of radiation is 4.5 mm above the expected value, and in December, the amount of radiation is 4.7 mm below the expected value. Thus we have $4.5 = a\sin(0.461011(6) + c)$ and $-4.7 = a\sin(0.461011(12) + c)$. We solve for a in the first equation $a = \frac{4.5}{\sin(0.461011(6) + c)}$ and substitute this into the second equation:

$$-4.7 = \frac{4.5}{\sin(0.461011(6) + c)}\sin(0.461011(12) + c).$$

Using technology, we find $c \approx 2.24801$ and $a \approx -4.71284$.

$E(t) = -4.71284\sin(0.461011t + 2.24801)$ mm per day, where t is the month of the year.

c. $E(t) = -4.71284\sin(0.461011t + 2.24801) + 12.5$ mm/day where t is the month of the year

d. In March the amount is $f(3) \approx 14.7$ mm per day, and in September the amount is $f(6) \approx 12.0$ mm per day. The model over-predicts by 2.7 mm per day in March and slightly underpredicts by 0.5 mm per day in September.

Chapter 8 Concept Review

1. a. The relative risk of having a car accident is changing with respect to blood alcohol level at a rate that is proportional to the risk of having a car accident at certain blood alcohol level.

b. Solve by separation of variables.

$$\frac{1}{R}dR = kdb$$

$$\int \frac{1}{R}dR = \int kdb$$

$$\ln R + c_1 = kb + c_2 \text{ (Because the risk is always positive, we omit the absolute value.)}$$

$$\ln R = kb + C$$

$$R = e^{kb+C}$$

$$R = ae^{kb}$$

$$R(b) = ae^{kb}$$

c. $R(0) = a = 1$ and $R(0.14) = ae^{0.14k} = 20$. Solving these equations, we get $a = 1$ and $k = \frac{\ln 20}{0.14} \approx 21.398$.

$R(b) = e^{21.398b}$ percent, where b is the proportion of alcohol in the blood stream

d. A certain occurrence corresponds to a relative risk of 100%, so $R = 100$. We solve $e^{21.398b} = 100$ for b:

$$e^{21.398b} = 100$$

$$21.398b = \ln 100$$

$$b = \frac{\ln 100}{21.398}$$

$$\approx 0.215$$

Thus according to the model, a crash is certain to occur when the blood alcohol level is approximately 21.5%.

e.

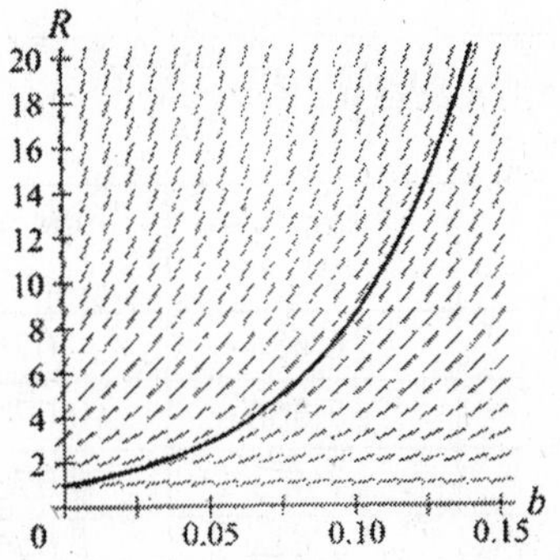

R
20
18
16
14
12
10
8
6
4
2
0
0.05
0.10
0.15
b

2. a. $\frac{dP}{dx} = 0.001175P(16.396 - P)$ million people per year x years after 1800

b. This differential equation has a logistic function as its solution.

$P(x) = \dfrac{16.396}{1+Ae^{-0.001175(16.396)x}} = \dfrac{16.396}{1+Ae^{-0.019265x}}$ million people is the population of Ireland x years after 1800

c. We use the fact that $P(-20) = 4.0$, to solve for A:

$$\frac{16.396}{1+Ae^{0.385306}} = 4$$

$$16.396 = 4(1+Ae^{0.385306})$$

$$12.396 = 4Ae^{0.385306}$$

$$A = \frac{12.396}{4e^{0.385306}} \approx 2.108$$

Thus $P(x) = \dfrac{16.396}{1+2.108e^{-0.019265x}}$ million people is the population of Ireland x years after 1800.

d. $P(40) \approx 8.3$ and $P(50) \approx 9.1$
We estimate that there were 8.3 million people in 1840 and 9.1 million people in 1850.

3. a.

x (years after 1800)	Estimate of $P(x)$ (million people)	Slope at x (million people per year)
–20	4	0.058
–10	4.583	0.064
0	5.219	0.069
10	5.904	0.073
20	6.632	0.076
30	7.393	0.078
40	**8.175**	0.079
50	**8.965**	

We estimate that $P(40) \approx 8.18$ million people; $P(50) \approx 8.97$ million people.

b.

x	Estimate of $P(x)$	Slope at x	x	Estimate of $P(x)$	Slope at x
–20	4	0.058	20	6.684	0.076
–15	4.291	0.061	25	7.066	0.077
–10	4.596	0.064	30	7.453	0.078
–5	4.915	0.066	35	7.844	0.079
0	5.247	0.069	**40**	**8.239**	0.079
5	5.590	0.071	45	8.633	0.079
10	5.945	0.073	**50**	**9.027**	
15	6.310	0.075			

We estimate that $P(40) \approx 8.24$ million people and $P(50) \approx 9.03$ million people.

c. The 1940 estimates in this question are 0.12 and 0.06 million people less than population found in the Activity 2. The 1950 estimates are 0.13 and 0.07 million people less than the Activity 2 answer.

4. a. $\frac{dQ}{dx} = -0.008307Q(7.154 - Q)$ million people per year, where the population is $P(x) = Q(x) + 4.4$ million people and x is the number of years since 1800

b. This differential equation has a logistic function as its solution.

$Q(x) = \frac{7.154}{1 + Ae^{0.008307(7.154)x}} = \frac{7.154}{1 + Ae^{0.059428x}}$ million people is the population of Ireland x years after 1800

Because $Q(100) = 4.5$, we have the equation $\frac{7.154}{1 + Ae^{0.59428}} = 0.1$. Solving this equation for A (as illustrated in Activity 2 part *c*), we get $A = \frac{7.054}{0.1e^{5.9428278}} \approx 0.185$.

$Q(x) = \frac{7.154}{1 + 0.185e^{0.059428x}}$ million people, where x is the number of years since 1800

c.

$$P(x) = \begin{cases} \frac{16.396}{1 + 2.108e^{-0.019265x}} \text{ million people} & \text{when } x \le 40 \\ \frac{7.154}{1 + 0.185e^{0.059428x}} + 4.4 \text{ million people} & \text{when } x \ge 50 \end{cases}$$

is the population of Ireland where x is the number of years since 1800

d. Using the model in part *c*, $P(50) \approx 6.0$

We estimate that there were 6.0 million people in 1850. This answer is significantly smaller than the one found in part *d* of Activity 2.

Chapter 9
Ingredients of Multivariable Change: Models, Graphs, Rates

Section 9.1 Multivariable Functions and Contour Graphs

1. a. $P(1.2, s)$ is the profit in dollars from the sale of a yard of fabric as a function of s, the selling price per yard, when the production cost is \$1.20 per yard.

   b. $P(c, 4.5)$ is the profit in dollars from the sale of a yard of fabric as a function of c, the production cost per yard, when the selling price is \$4.50 per yard.

   c. When the production cost is \$1.20 per yard and the selling price is \$4.50 per yard, the profit is \$3.00 for each yard sold.

   d.

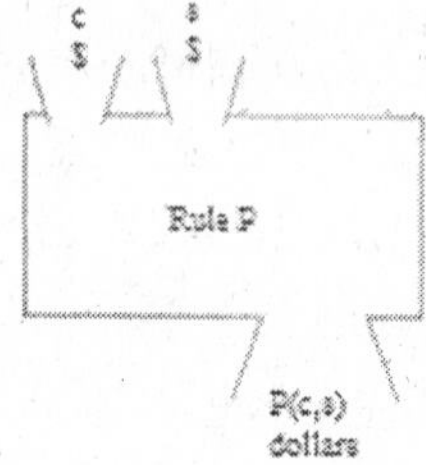

3. a. $P(100{,}000, m)$ is the probability of the senator voting in favor of a the bill as a function of the amount m, in millions of dollars, invested by the tobacco industry lobbying against the bill, when the senator receives 100,000 letters supporting the bill.

   b. $P(l, 53)$ is the probability of the senator voting in favor of the bill as a function of l, the number of letters supporting the bill received by the senator, when the tobacco industry spends \$53 million lobbying against the bill.

   c.

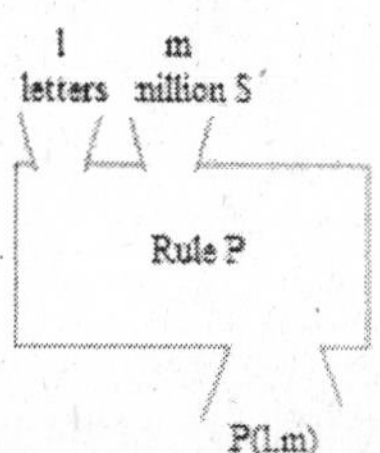

5.

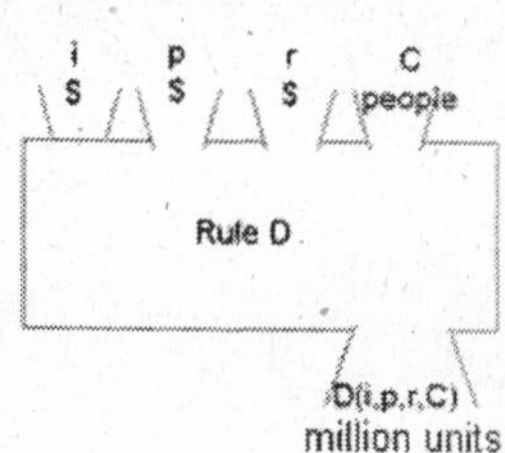

7.

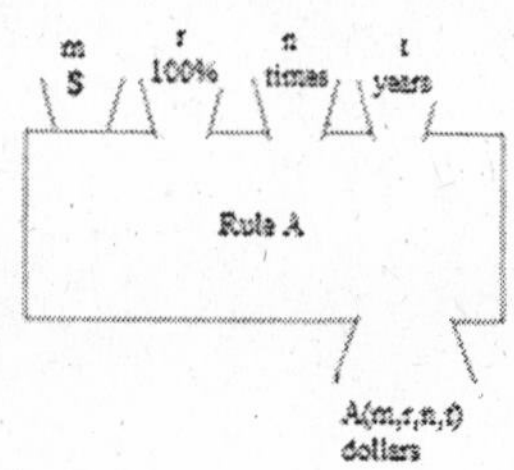

9.

Air temperature (°F)	Relative humidity (%)													
	40	45	50	55	60	65	70	75	80	85	90	95	100	
110	135													
108	130	137												
106	124	130	137											
104	119	124	130	137										
102	114	119	124	130	137									
100	109	113	118	123	129	136								
98	105	108	113	117	122	128	134							
96	101	104	107	111	116	121	126	132						
94	97	100	103	106	110	114	119	124	129	135				
92	94	96	98	101	104	108	112	116	121	126	131			
90	91	92	94	97	99	102	106	109	113	117	122	126	131	130
88	88	89	91	93	95	97	100	103	106	109	113	117	121	
86	85	86	88	89	91	93	95	97	99	102	105	108	111	
84	83	84	85	86	87	89	90	92	94	96	98	100	102	105
82	81	82	83	83	84	85	86	87	88	90	91	93	94	
80	80	80	81	81	82	82	83	83	84	85	85	86	87	90

11. a. There will be 11.97 hours of daylight.

b. There will be 12.31 hours of daylight.

c. Answers will vary. For example, if your college is at the 45th parallel north, it will receive 9.19 hours of daylight each day during January.

d.

Latitude North South		Month												
		Jan Jul	Feb Aug	Mar Sep	Apr Oct	May Nov	Jun Dec	Jul Jan	Aug Feb	Sep Mar	Oct Apr	Nov May	Dec Jun	
0	12	12.12	12.12	12.12	12.12	12.12	12.12	12.12	12.10	12.11	12.12	12.12	12.12	12
5		11.87	11.96	12.08	12.22	12.35	12.41	12.38	12.28	12.16	12.02	11.90	11.83	
10		11.61	11.81	12.06	12.35	12.57	12.70	12.64	12.45	12.17	11.91	11.67	11.55	
15		11.34	11.66	12.04	12.47	12.82	13.00	12.92	12.62	12.22	11.81	11.44	11.25	11
20	11	11.07	11.50	12.01	12.60	13.07	13.32	13.22	12.81	12.26	11.70	11.20	10.94	
25		10.78	11.33	11.97	12.74	13.34	13.66	13.53	13.02	12.31	11.58	10.94	10.62	
30		10.45	11.14	11.97	12.88	13.65	14.05	13.88	13.23	12.35	11.47	10.67	10.26	10
35	10	10.09	10.95	11.95	13.06	13.98	14.47	14.27	13.47	12.42	11.33	10.36	9.86	
40		9.68	10.71	11.91	13.25	14.36	14.96	14.71	13.76	12.48	11.18	10.00	9.39	9
45	9	9.19	10.45	11.87	13.48	14.82	15.55	15.25	14.09	12.55	11.01	9.60	8.85	
50		8.61	10.13	11.84	13.78	15.38	16.29	15.91	14.48	12.66	10.80	9.07	8.17	
55		7.83	9.73	11.79	14.10	16.14	17.28	16.78	14.99	12.76	10.55	8.45	7.28	
60		6.79	9.21	11.74	14.62	17.10	18.70	18.01	15.67	12.92	10.22	7.60	6.04	

18 17 16 15 14 13

13.

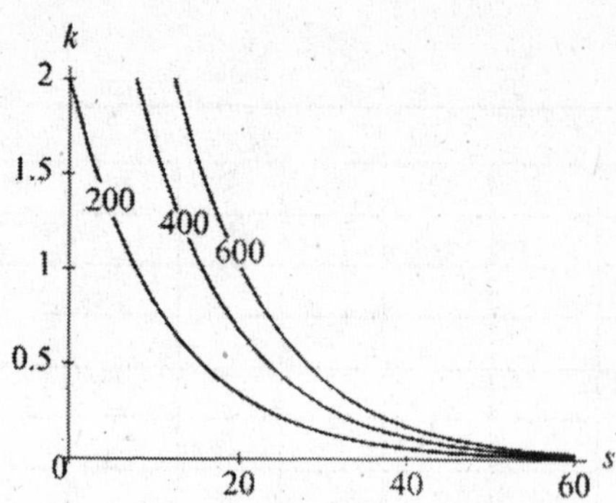

15.

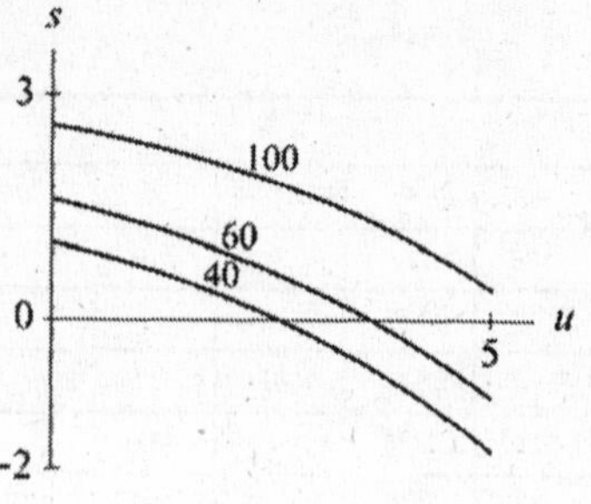

17.

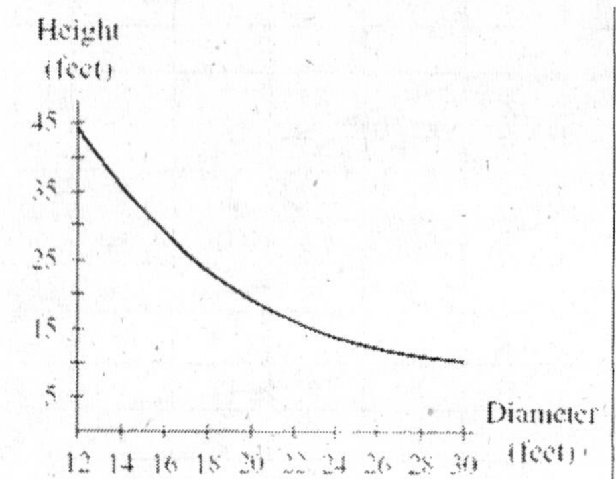

19.a. $P(c, s) = P(c, 100, 10, s) = 0.175c + 0.027s^2 - 0.730s + 108.958$ for a supermarket with s thousand square feet of sales space and a customer base with a per capita income of $\$c$ thousand

b.

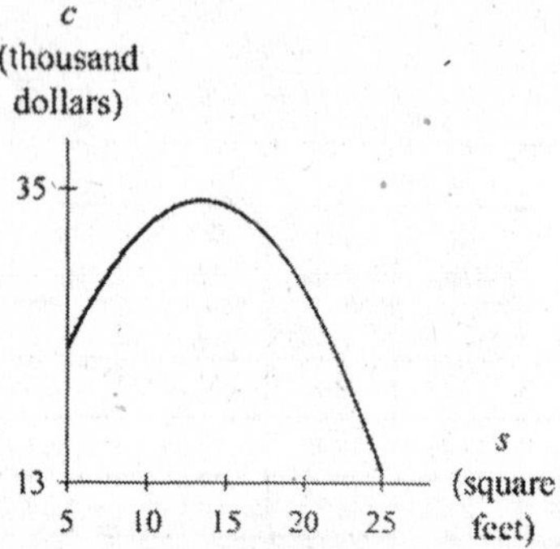

21. a. For example, the BMI for a person who is 5'8" and 150 pounds is 22.8 points.

b.

Weight (pounds)	Height (inches)						
	60	62	64	66	68	70	72
90	17.6	16.5	15.4	14.5	13.7	12.9	12.2
100	19.5	18.3	17.2	16.1	15.2	14.3	13.6
110	21.5	20.1	18.9	17.8	16.7	15.8	14.9
120	23.4	21.9	20.6	19.4	18.2	17.2	16.3
130	25.4	23.8	22.3	21.0	19.8	18.7	17.6
140	27.3	25.6	24.0	22.6	21.3	20.1	19.0
150	29.3	27.4	25.7	24.2	22.8	21.5	20.3
160	31.2	29.3	27.5	25.8	24.3	23.0	21.7
170	33.2	31.1	29.2	27.4	25.8	24.4	23.1
180	35.2	32.9	30.9	29.1	27.4	25.8	24.4
190	37.1	34.8	32.6	30.7	28.9	27.3	25.8
200	39.1	36.6	34.3	32.3	30.4	28.7	27.1

15 20 25 35 30

c. Replace the output with a constant K and solve for one variable in terms of the other. We choose to solve for w.

$$K = \frac{0.4536w}{0.00064516h^2}$$

$$w = \frac{K(0.00064516)h^2}{0.4536}$$ pounds, where h is the height in inches and K is the BMI

d.

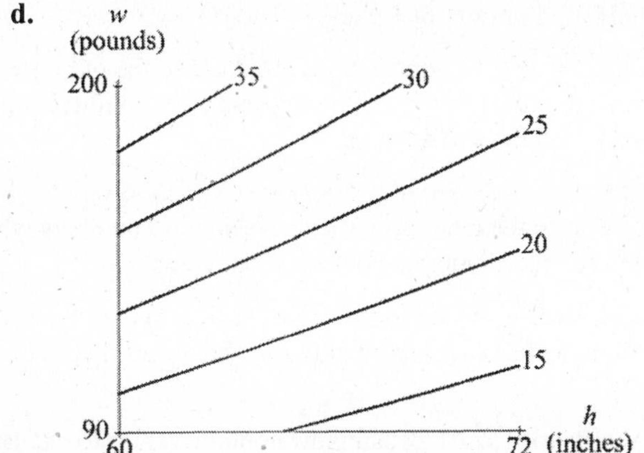

This graph is the upside-down mirror reflection of the graph in part *b*. It is also more accurately drawn.

23. a. Solve for one of the variables in $10.65 + 1.13w + 1.04s - 5.83ws = K$ (we show both).

$$s(1.04 - 5.83w) = K - 10.65 - 1.13w$$
$$s = \frac{K - 10.65 - 1.13w}{1.04 - 5.83w}$$

or

$$w(1.13 - 5.83s) = K - 10.65 - 1.04s$$
$$w = \frac{K - 10.65 - 1.04s}{1.13 - 5.83s}$$

b.

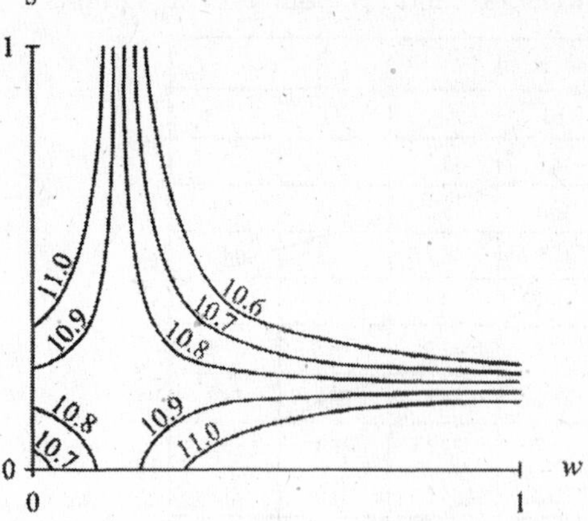

25. a. The center of the 184 thousand-contour curve corresponds to approximately 10 thousand tons of lower fat cheese and 55 thousand tons of regular cheese since is approximately (10, 55). This point corresponds to a maximum because the contour curves decrease in every direction away from the point.

b. The contour graph shows that the maximum revenue is greater than 184 million guilders but less than 214 million guilders (or else we would see the 214-contour curve). From the three-dimensional graph, it appears that the maximum revenue is near 200 million guilders. One possible approximation is 190 million guilders.

27. a. Estimates will vary. The input values are $P \approx 10$ hours and $H \approx 70\%$, and the output value is approximately 12.5 days. When *C. grandis* is exposed to 70% relative humidity and 10 hours of light, it takes approximately 12.5 days to develop.

b. Estimates will vary. The input values are $P \approx 10$ hours and $H \approx 60\%$, and the output value is approximately 11.5 days. When *C. grandis* is exposed to 60% relative humidity and 10 hours of light, it takes approximately 11.5 days to develop.

29. a. $f(x, y)$ decreases when y increases because the contour curves have smaller numbers when y increases from 2 and x is 1.5; therefore, $f(x, y)$ increases when y decreases.

b. Because the contour curves are more closely spaced to the left of (2.5, 2.5) than they are directly below that point, the function decreases more quickly as x decreases than it does as y decreases.

c. The change is greater when (2, 2) shifts to (1, 2.5), causing the contour values to change from approximately 21 to approximately 14, than it is when (1, 0) shifts to (4, 1), causing virtually no change in contour values.

31. a. The three dimensional graph has two peaks the approximately the same height —one around (0.7, 0, 0.25) and one around (–0.7, 0. 0.25) —separated by a valley.

b. The point (0.4, 0.4) lies between the 0 and –0.05 contours. The point (0. 0.3) lies near the –0.1 contour. Thus the descent is greater from (0.7, 0.1) to (0, 0.3).

c. Moving up from (0, 0.1), we encounter increasingly negative contours. Moving right from (0, 0.1) we encounter increasingly positive contours. Thus the function output increases as x increases from (0, 0.1).

d. The point (–0.2, –0.3) lies near the –0.05 contour. Thus we are looking for a point lying on the –0.05 + 0.15 = 0.1 contour. There are infinitely many such points. Two possibilities are (–0.7, 0.38) and (0.94, 0).

33. a.

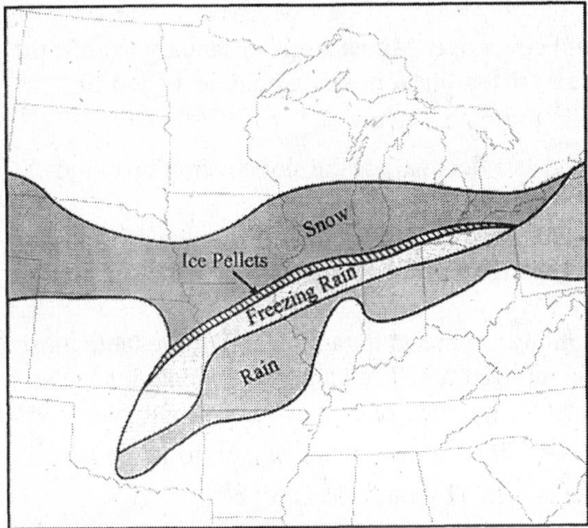

b. The temperature dropped 36°F.

35. To locate a relative maximum on a contour graph look for contour curves that form a simple, closed curve (or if completed appear to form a closed curve). Estimate the point in the center of this closed curve. Moving away from the "center" check that contour levels are decreasing. If the levels are decreasing by a constant amount d, the maximum value at the point can estimated between the largest indicated contour value k and $k+d$.

On a table of data, locate a number whose eight surrounding numbers are all less than it. It may help to spot this number if you first quickly sketch a couple of contour curves. Once you have located a number that appears to be a maximum value, you describe its location by following its column and row to the edges of the table to determine the input values that produce that maximum.

Keep in mind that for a contour graph as well as a table, there may exist more than one relative maximum.

Section 9.2 Cross-sectional Models and Rates of Change

1. **a.** air temperature

   b. row

   c. $A(p, 95) = 0.0238p^2 - 2.3455p + 151.31$ °F when the dew point is p°F.

3. **a.** At least $\frac{8}{10}$ of the sky is covered by clouds 60% of the time.

   b. The time of day is given as 9 A.M. so the variable to be held constant is *hour of the day.* The input variable for the cross-sectional model is *fraction of sky covered.*

   c. column

   d. A plot of the frequency of cloud cover over Minneapolis in January against the fraction of sky covered at 9:00 a.m. indicates that a linear model would be a good fit.

   $C(9, f) = -1.651f^3 + 2.686f^2 - 1.597f + 1.019$

   where f is the fraction (expressed as a decimal) of the sky covered by clouds.

5. **a.** no; In order to model monthly payments for a 52-month loan, the table must include a column for a 52-month loan given different interest rates. The table shown does not include such a column.

   b. yes; row; In order to model monthly payments for a loan at 9%, the table must include a row for 9% interest given different lifetimes. The table shown includes such a row.

   c. A possible cross-sectional model for monthly payments for a \$1000 loan at 9% is $p(m,9) = 0.017m^2 - 2.091m + 85.977$ dollars when m months is the length of the loan (m between 24 and 60). Payments on a 52 month loan will be $p(52,9) = 0.017(52)^2 - 2.091(52) + 85.97 \approx 22.65$ dollars each month.

7. **a.** Using price per pound above \$1.50 as input and the data in the row labeled 4 as output, we choose a quadratic as a possible model. A model for the per capita consumption of peaches by people living in households with \$40,000 yearly income is

   $C(p,4) = 0.893p^2 - 1.304p + 7.811$ pounds per person per year where p and 1.50 dollars per pound is the price of peaches.

   b. When peaches sell for \$1.55 per pound, $p = 1.55 - 1.50 = 0.05$. The per capita consumption at this price by someone in a household with \$40,000 yearly income is $C(0.05, 4) = 0.893(0.05)^2 - 1.304(0.05) + 7.811 \approx 7.7$ pounds per person per year.

9. **a.** Since there is a column for \$1.00 coke products but not a row for 90 cent Pepsi products, we must hold the cost of Coke products constant.

   b. We use the column for \$1.00 Coke products.

   c. Using cost of Pepsi products (aligned to cents, i.e. 50, 75, 100, 125, 150) as input and the data from the \$1.00 Coke products column as output, we obtain the following linear model for daily sales of Coke products: $S(100, p) = 1.96p + 25$ cans when Pepsi products are sold for p cents per can. Using this model to estimate the number of Coke products sold at \$1.00 when Pepsi products sell for \$0.90, we calculate $S(100, 90) = 1.96(90) + 25 \approx 201$ cans.

11. a. 3.88 million people

b. See figure on text page A-71.

c. We find a cross-sectional model

$P(2005,\ a)=\begin{cases}-0.0506a+5.036 & \text{when } 15\le a\le 30\\ -0.0064a^2+0.548a-7.240 & \text{when } 30\le a\le 50\end{cases}$ million people of age a in 2005.

$\left.\frac{\partial}{\partial a}P(2005,\ a)\right|_{a=35}=-0.0065\cdot 2(35)+0.548\approx 0.1$ million people per year of age a.

d. We find a cross-sectional model

$P(y,\ 50)=-\left(1.778\cdot 10^{-4}\right)y^3+0.004y^2+0.106y+2.405$ million people of age 50 y years after 1990.

$\left.\frac{\partial}{\partial y}P(y,\ 50)\right|_{y=30}=-\left(1.778\cdot 10^{-4}\right)\left(3\cdot(30)^2\right)+0.004(2\cdot(30))+0.106\approx -0.2$ million people per year

e. The answer to part *c* shows that moving down off the 400-contour curve at (2005, 35) will result in a an increase of output. The answer to part *d* shows that moving right off the 400-contour curve at (2020, 50) will result in a decrease of output.

13. a. -0.35 kg/day (loss of 0.35 kilograms per day)

b. W(68,t) represents the cross-sectional model of the first column in the table.
$W(68,t)=-0.00009t^3+0.003t^2-0.013t+0.572$ kg/day where t°C is the temperature.
A pig weighing 68 kg will (on average) gain/loose W(68,*t*) kilograms during a day when the temperature is t°C.

c. The graph of W(68,*t*) is concave up, increasing until approximately $t=10$. After $t=10$, the graph is concave down. It reaches a maximum just after $t=21.1$. A pig weighing 68 kilograms will gain the most weight when the temperature is near 21.1°C. For temperatures in excess of 21.1°C, the weight gain diminishes more and more rapidly until weight is actually lost at temperatures in excess of approximately 37.5°C.

15. a. Approximately $0.52°F$ per $°F$

b. First, determine a cross-sectional model:
$A(p,\ 85)=0.0194p^2-2.0052p+135.832$°F when the dew point is p°F.

Second, determine the derivative of this model: $\frac{d(p,85)}{dp} = 0.0388p - 2.0052$

Third, evaluate the derivative for $p = 65°F$

17. a. $A(14{,}000, r) = 14{,}000r^2 + 28{,}000r + 14{,}000$ dollars
where 100%r is the annual percentage yield (i.e., r is in decimals).

b. $\frac{dA(14{,}000, r)}{dr} = 28{,}000r + 28{,}000$ dollars per 100 percentage points

$\frac{dA(14{,}000, 0.127)}{dr} = 28{,}000(0.127) + 28{,}000 = \$31{,}556 \;/\; 100$ percentage points

c. $A(14{,}000, r) = 1.4r^2 + 280r + 1400$ dollars, where r% is the annual percentage yield

The magnitudes of the coefficients are reduced in this model, the r^2 coefficient by a factor of 10,000, the r coefficient by a factor of 100, and the constant by a factor of 10.

d. $\frac{dA(14{,}000, r)}{dr} = 2.8r + 280$ dollars per percentage point

$\frac{dA(14{,}000, 12.7)}{dr} = 2.8(12.7) + 280 = \$315.56$ per percentage point

The derivative tells us approximately how much the output will change when the input increases by one unit. If the input is a percentage expressed as a decimal, then an increase in one unit corresponds to 100 percentage points. For example, if $r = 0.127$ and is increased by 1 to $r = 1.127$ the corresponding percentages are 12.7% and 112.7%.

This answer is equivalent to the one found in part *b.*

19. Derivatives of cross-sectional models of a three-dimensional function can be used to pin-point optimal points on those cross-sections. These optimal points in turn may e used to estimate the point at which a critical point may appear on the three-dimensional function.

Section 9.3 Partial Rates of Change

1. $\frac{\partial W}{\partial h}$ pounds per inch

3. $\left.\frac{\partial T}{\partial t}\right|_{g=23}$ °F per degree of latitude

5. $\left.\frac{\partial R}{\partial c}\right|_{b=2}$ dollars per cow for $c = 100$

7. a. $\left.\frac{\partial P}{\partial m}\right|_{l\,=\,100{,}000}$ is the rate of change of the probability that the senator will vote for the bill with respect to the amount spent by the tobacco industry on lobbying when the senator receives 100,000 letters in opposition to the bill. We expect this rate of change to be negative because if the number of letters is constant but lobbying funding against the bill increases, the probability that the senator votes for the bill is likely to decline.

b. $\left.\frac{\partial P}{\partial l}\right|_{m\,=\,53}$ is the rate of change of the probability that the senator votes for the bill with respect to the number of letters received when \$53 million is spent on lobbying efforts. We expect this rate of chance to be positive because if the number of letters increases (while lobbying funding remains constant) the probability of the senator voting in favor of the bill is likely to increase.

9. a. $\frac{\partial f}{\partial x} = 3(2)x + 5y(1) + 0$

$= 6x + 5x$

b. $\frac{\partial f}{\partial y} = 0 + 5x(1) + 2(3y^2)$

$= 5x + 6y^2$

c. $\left.\frac{\partial f}{\partial x}\right|_{y\,=\,7} = 6x + 5(7)$

$= 6x + 35$

11. a. $\frac{\partial f}{\partial x} = 5(3x^2) + 3(2x)y^3 + 9(1)y + 14(1) + 0$

$= 15x^2 + 6xy^3 + 9y + 14$

b. $\frac{\partial f}{\partial y} = 0 + 3x^2(3y^2) + 9x(1) + 0 + 0$

$= 9x^2y^2 + 9x$

c. $\left.\frac{\partial f}{\partial x}\right|_{y=2} = 15x^2 + 6x(2)^3 + 9(2) + 14$

$= 15x^2 + 48x + 32$

13. a. $M_t = s\left(\frac{1}{t}\right) + 0 + 0 = \frac{s}{t}$

b. $M_s = \ln t + 3.75$

c. $\left.M_s\right|_{t\,=\,3} = \ln 3 + 3.75$

15. a. $\frac{\partial h}{\partial s} = \frac{1}{t} - 2(st - tr)^1(t - 0) = \frac{1}{t} - 2t(st - tr)$

b. $\frac{\partial h}{\partial t} = -1(s)t^{-2} + \frac{1}{r} - 2(st - tr)^1(s - r) = \frac{-s}{t^2} + \frac{1}{r} - 2(st - tr)(s - r)$

c. $\frac{\partial h}{\partial r} = 0 + -1(t)r^{-2} - 2(st - tr)^1(0 - t) = \frac{-t}{r^2} + 2t(st - tr)$

d. $\left.\frac{\partial h}{\partial r}\right|_{(s,\,t,\,r)=(1,\,2,\,-1)} = \frac{-2}{(-1)^2} + 2(2)[1(2) - 2(-1)] = -2 + 4(4) = 14$

17. a. $f_x = 2y + 8(2x)y^3 + 0 + 0$

$= 2y + 16xy^3$

b. $f_y = 2x + 8x^2(3y^2) + 5(\ln e^2)e^{2y} + 0$

$= 2x + 24x^2y^2 + 10e^{2y}$ note: $\ln e^2 = 2$

c. $f_{xx} = 0 + 16y^3$

$= 16y^3$

$f_{xy} = 2 + 16x(3y^2)$

$= 2 + 48xy^2$

$f_{yx} = 2 + 24(2x)y^2 + 0$

$= 2 + 48xy^2$

$f_{yy} = 0 + 24x^2 92y) + 10(\ln e^2)e^{2y}$

$= 48x^2y + 20e^{2y}$

d. $\begin{bmatrix} 16y^3 & 2 + 48xy^2 \\ 2 + 48xy^2 & 48x^2y + 20e^{2y} \end{bmatrix}$

19.

$$\begin{array}{cc} & \begin{array}{cc} x & y \end{array} \\ \begin{array}{c} x \\ \\ y \end{array} & \begin{bmatrix} \frac{-2y}{x^3} & \frac{-1}{y^2} + \frac{1}{x^2} \\ & \\ \frac{-1}{y^2} + \frac{1}{x^2} & \frac{2x}{y^3} \end{bmatrix} \end{array}$$

21.
$$\begin{array}{c|cc} & x & y \\ \hline x & 4e^{2x-3y} & -6e^{2x-3y} \\ y & -6e^{2x-3y} & 9e^{2x-3y} \end{array}$$

23. a. $A(14{,}000,\ r) = 14{,}000(1+r)^2$ dollars. This function gives the value of an investment after 2 years when the APY is $100r\%$.

b. $\left.\dfrac{\partial A}{\partial r}\right|_{P=14{,}000} = \dfrac{dA(14{,}000,\ r)}{dr} = 28{,}000(1+r)$ dollars per 100 percentage points

$\left.\dfrac{\partial A}{\partial r}\right|_{(P,\ r)\ =\ (14{,}000,\ 0.127)} = \$31{,}556$ per 100 percentage points

The value of an investment of \$14,000 after 2 years in an account with A.P.R. of 12.7%, is increasing by \$315.56 per additional percentage point.

c. The slope of the line tangent to a graph of $A(14{,}000,\ r)$ at $r = 0.127$ is \$31,566 per 100 percentage points.

(*Note:* It is difficult to distinguish the graph of A from the tangent line in this figure because the graph of A appears nearly linear in this close-up view. This is an illustration of the principle of local linearity discussed in Chapter 1.)

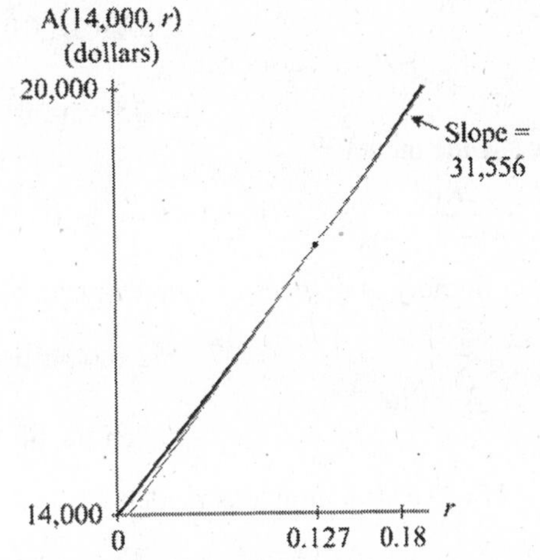

25. a. A cross-sectional model for the amount of UVA radiation with respect to the latitude for the month of March is $R(\text{March}, l) = -0.0073l^2 - 0.0346l + 49.7492$ watts/m^2/month where l is the number of degrees of latitude north of the equator (negative values of l correspond to locations south of the equator).

An alternative model a sine model: $R(\text{March}, l) = 35.9104\sin(-0.0232l + 1.5088) + 15.5463$ watts/m^2/month where l is the number of degrees of latitude north of the equator.

A cross-sectional model for the amount of UVA radiation with respect to the month at 50° south of the equator is $R(m, -50) = 27.1052\sin(0.4846m + 1.7070) + 32.9106$ watts/m^2/month where m is the number of months since the beginning of the year.

An alternative model is the piecewise continuous model

$$R(m,-50^\circ)=\begin{cases}0.389m^3-3.321m^2-3.425m+61.857 & \text{when } 0\le m<7\\ \text{watts per square meter per month} & \\ -0.565m^3+16.008m^2-137.689m+381.183 & \text{when } 7\le m\le 12\\ \text{watts per square meter per month} & \end{cases}$$

b. Quadratic model:

$$\left.\frac{\partial R}{\partial l}\right|_{m=\text{March}}=-0.0146l-0.0342 \text{ watts/m}^2\text{/month/degree of latitude}$$

where l is the number of degrees of latitude north of the equator.

$$\left.\frac{\partial R}{\partial l}\right|_{(m,l)=(\text{March},-50)}=-0.0146(-50)-0.0342\approx 0.70 \text{ watt/m}^2\text{/month/degree of latitude}$$

Sine model:

$$\left.\frac{\partial R}{\partial l}\right|_{m=\text{March}}=35.2303(-0.0236)\cos(-0.0236l+1.5069) \text{ watts/m}^2\text{/month/degree of latitude}$$

where l is the number of degrees of latitude north of the equator.

$$\left.\frac{\partial R}{\partial l}\right|_{(m,l)=(\text{March},-50)}=35.2303(-0.0236)\cos(-0.0236(-50)+1.5069)$$

$$\approx 0.75 \text{ watt/m}^2\text{/month/degree of latitude}$$

c. Sine model:

$$\left.\frac{\partial R}{\partial m}\right|_{l=-50}=27.1052(0.4846)\cos(0.4846m+1.7070) \text{ watts/m}^2\text{/month/month}$$ where m is the number of degrees of latitude north of the equator.

$$\left.\frac{\partial R}{\partial m}\right|_{(m,l)=(3,-50)}=27.1052(0.4846)\cos(0.4846(3)+1.7070)$$

$$\approx -13.13 \text{ watts/m}^2\text{/month/month}$$

Piecewise continuous model:

$$\left.\frac{\partial R}{\partial m}\right|_{l=-50}=\begin{cases}3(0.389)m^2-2(3.321)m-3.425 & \text{when } 0\le m<7\\ \text{watts per square meter per month}^2 & \\ 3(-0.565)m^2+2(16.008)m-137.689 & \text{when } 7< m\le 12\\ \text{watts per square meter per month}^2 & \end{cases}$$

$$\left.\frac{\partial R}{\partial m}\right|_{(m,l)=(3,-50)}=3(0.389)3^2-2(3.321)3-3.425\approx -12.85 \text{ watts/m}^2\text{/month/month}$$

d.

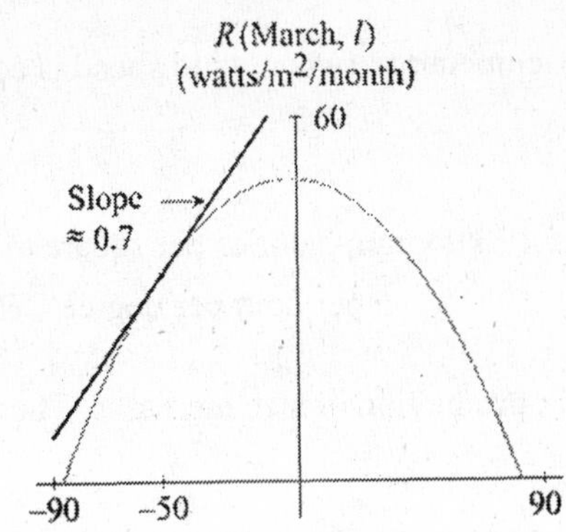

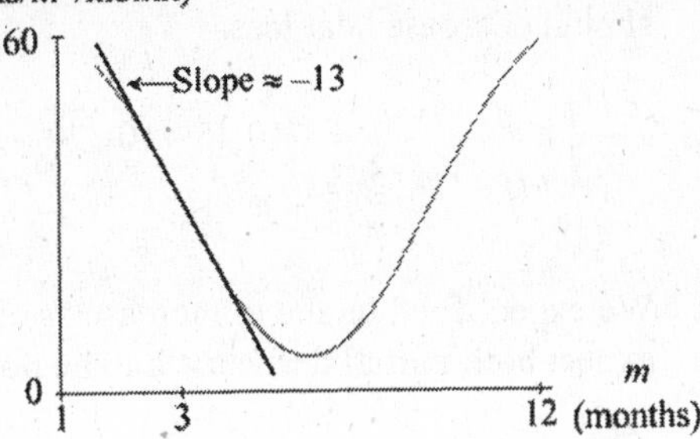

27. a. A logarithmic model is $C(0.2,i)=4.805+2.002\ln i$ peaches per person, at a price of \$1.70 per pound when the yearly income is i tens of thousands of dollars.

$$\left.\frac{\partial C}{\partial i}\right|_{(3)}=0.67 \text{ peaches per person per ten thousand dollars}$$

b. An exponential model is $C(p,3)=7.177(0.896^{p})$ peaches per person, at a yearly income of \$30,000 and a price per pound of (p +1.5) \$.

$$\left.\frac{\partial C}{\partial p}\right|_{(0.2)}=-0.77 \text{ peaches per person per \$ per pound}$$

c. Using the provided model to answer part *a* results in $\frac{\partial C}{\partial i}\approx 0.8$ pound/person/year per thousand dollars of income

Using the provided model to answer part *b* results in $\frac{\partial C}{\partial p}\approx -0.82$ pound/person/year per dollar per pound

d. Parts (a) and (b) were easier to use since they were single variable functions. I would expect that part (c) was more accurate since we were given the appropriate model.

29. a. $\frac{\partial H}{\partial t}=(10.45+10\sqrt{v}-v)(-1)=-10.45-10\sqrt{v}+v$ kilogram-calories per square meter per hour per degree Celsius

$\frac{\partial H}{\partial v}=(\frac{1}{2}v^{-1/2}-1)(33-t)=(33-t)\left(\frac{5}{\sqrt{v}}-1\right)$ kilogram-calories per square meter per hour per meter per second

b. $\frac{\partial H}{\partial v}$ should be positive because an increase in wind speed (when temperature is constant) should increase heat loss.

c. $\left.\frac{\partial H}{\partial v}\right|_{(v,\,t)\,=\,(20,\,12)}=(33-12)\left(\frac{5}{\sqrt{20}}-1\right)\approx 2.48$ kilogram-calories per square meter per hour per meter per second

d. $\frac{\partial H}{\partial t}$ should be negative because an increase in temperature (when wind speed is constant) should decrease heat loss.

e. $\left.\frac{\partial H}{\partial t}\right|_{(v,\,t)\,=\,(20,\,12)} = -10.45 - 10\sqrt{20} + 20 \approx -35.17$ kilogram-calories per square meter per hour per degree Celsius

31. a. We expect food intake to increase as either milk production or size increases. Therefore, we expect both partial derivatives to be positive.

b. $\frac{\partial I}{\partial s} = -1.244 + 0.1794s + 0.21491m$ kilograms per day per unit of size index

This equation is the rate of change of the amount of organic matter eaten with respect to the size of the cow (when the amount of milk produced is constant).

$\frac{\partial I}{\partial m} = -0.20988 + 0.071894m + 0.214915s$ kilograms per day per kilogram of milk

This equation is the rate of change of the amount of organic matter eaten with respect to the amount of milk produced (when the size of the cow is constant).

c. $\left.\frac{\partial I}{\partial m}\right|_{(s,\,m)\,=\,(2,\,6)} = -0.20988 + 0.071894(6) + 0.214915(2) \approx 0.65$ kilogram per day per kilogram of milk per day

d. $\left.\frac{\partial I}{\partial s}\right|_{(s,\,m)\,=\,(2,\,6)} = -1.244 + 0.1794(2) + 0.21491(6) \approx 0.40$ kilogram per day per unit of size index

e. $\frac{\partial^2 I}{\partial s^2} = 0.1794;\quad \frac{\partial^2 I}{\partial m^2} = 0.071894;\quad \frac{\partial^2 I}{\partial m \partial s} = \frac{\partial^2 I}{\partial s \partial m} = 0.214915$

$$\begin{array}{c} \\ s \\ m \end{array}\begin{array}{c} \begin{array}{cc} s \quad\quad & \quad m \end{array} \\ \begin{bmatrix} 0.1794 & 0.214915 \\ 0.214915 & 0.071894 \end{bmatrix} \end{array}$$

Because all second partials are positive, we know that the rates of change in the s and m directions increase as both s and m increase. This indicates that the surface is concave up in the s and m directions.

33. a. $\frac{\partial A}{\partial t} = 1000re^{rt}$ dollars per year and $\frac{\partial A}{\partial r} = 1000te^{rt}$ dollars per 100 percentage points

b. The second partials matrix is $\begin{array}{c} \\ t \\ r \end{array}\begin{array}{c} \begin{array}{cc} t \quad\quad\quad & \quad\quad r \end{array} \\ \begin{bmatrix} 1000r^2e^{rt} & 1000e^{rt}(rt+1) \\ 1000e^{rt}(rt+1) & 1000t^2e^{rt} \end{bmatrix} \end{array}$. Note that we use the Product Rule to find the mixed partials. Evaluating this matrix at $t = 30$ and $r = 0.047$, we have $\begin{bmatrix} 9.05 & 9871.25 \\ 9871.25 & 3{,}686{,}359.86 \end{bmatrix}$.

When a $1000 investment has been earning 4.7% compounded continuously for 30 years,

(1) the rate at which the amount is growing with respect to time is increasing with respect to time by \$9.05 per year per year.
(2) the rate at which the amount is growing with respect to time is increasing with respect to the interest rate by \$9871.25 per year per 100 percentage points.
(3) the rate at which the amount is growing with respect to the interest rate is increasing with respect to time by \$9871.25 per 100 percentage points per year.
(4) the rate at which the amount is growing with respect to the interest rate is increasing with respect to the rate by \$3,686,359.86 per 100 percentage points per 100 percentage points.

35. a. $\frac{\partial A}{\partial t} = (1+r)^t \ln(1+r)$ million dollars per year

b. $\frac{\partial A}{\partial r} = t(1+r)^{t-1}$ million dollars per 100 percentage points

c. When $t = 5$ and $r = 0.15$, $\frac{\partial A}{\partial t} \approx 0.28$ million dollars per year

d.

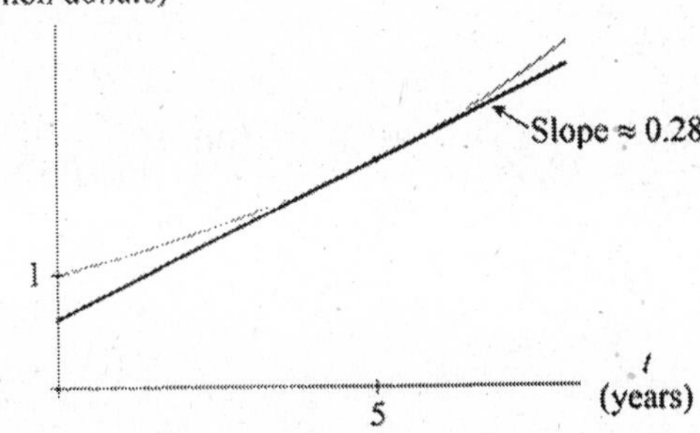

37. At a relative maximum, the partial derivatives must be equal to zero. At the instant of the high point, the output is neither increasing nor decreasing regardless of the input direction.

Section 9.4 Compensating for Change

1. $\frac{\partial f}{\partial x} = 30xy^3$, $\frac{\partial f}{\partial y} = 45x^2y^2$

$\frac{dx}{dy} = \frac{-f_y}{f_x} = \frac{-(45x^2y^2)}{30xy^3} = \frac{-3x}{2y}$

3. $\frac{\partial g}{\partial m} = \frac{59.3}{m} + 49n$, $\frac{\partial g}{\partial n} = 49m$

$\frac{dm}{dn} = \frac{-g_n}{g_m} = \frac{-49m}{\frac{59.3}{m} + 49n}$

5. $\frac{\partial g}{\partial x} = 1.05^y$ $\frac{\partial g}{\partial y} = x(\ln 1.05)1.05^y$

$\frac{dx}{dy} = \frac{-g_y}{g_x} = -x\ln 1.05$

When $y = 5$ and $g(x, y) = 100$, $x = \frac{100}{1.05^5} \approx 78.35$, so $\frac{dx}{dy} = -\left(\frac{100}{1.05^5}\right)\ln 1.05 \approx -3.82$.

Alternatively, $\frac{dy}{dx} \approx \frac{1}{-3.82} \approx -0.26$.

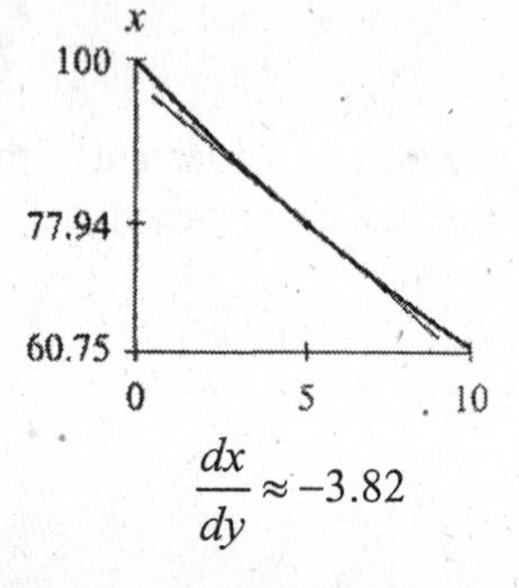

$\frac{dx}{dy} \approx -3.82$

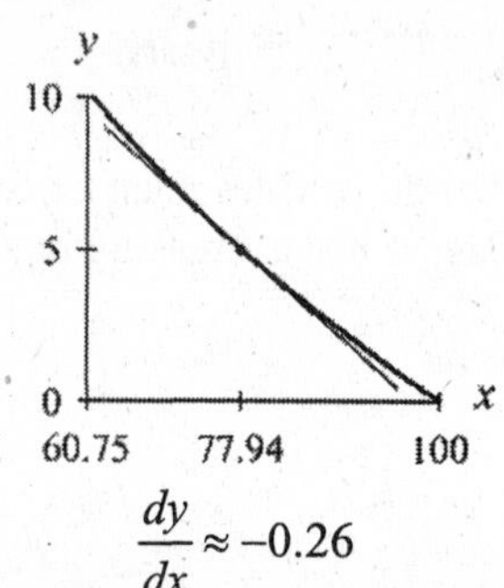

$\frac{dy}{dx} \approx -0.26$

7. $\frac{\partial f}{\partial a} = 5.6ab^3 - 3.6a$ $\frac{\partial f}{\partial b} = 8.4a^2b^2 + 12$

$\frac{da}{db} = \frac{-f_b}{f_a} = \frac{-(8.4a^2b^2 + 12)}{5.6ab^3 - 3.6a}$

When $b = 0.9$ and $f(a, b) = 15$, $a \approx 4.173$, so $\frac{da}{db} \approx -64.82$.

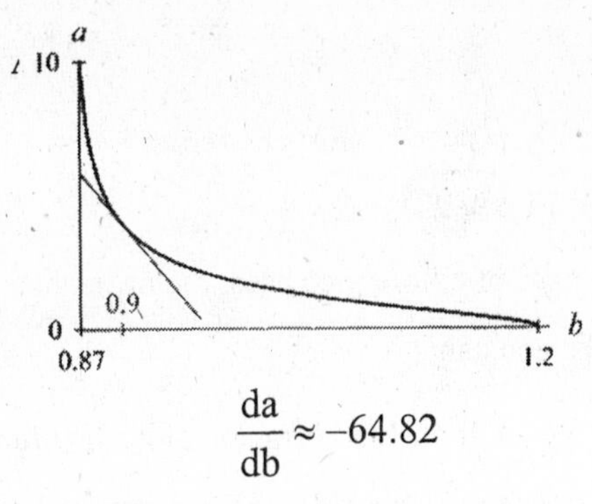

$$\frac{da}{db} \approx -64.82$$

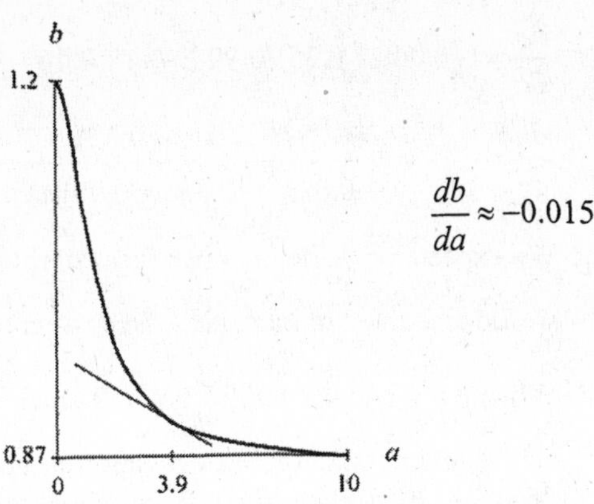

$$\frac{db}{da} \approx -0.015$$

9. $f(2, 1) = 21$

$\frac{\partial f}{\partial m} = 6m + 2n$, $\frac{\partial f}{\partial n} = 2m + 10n$

$\frac{dm}{dn} = \frac{-f_m}{f_n} = \frac{-(2m+10n)}{6m+2n}$

When $m = 2$, $n = 1$, and $\Delta m = 0.2$, $\frac{dm}{dn} = \frac{-14}{14} = -1$ and $\Delta n \approx \frac{dn}{dm}\Delta m = (-1)(0.2) = -0.2$.

The value of n should decrease be approximately 0.2 in order to compensate for an increase of 0.2 in m.

11. $f(3.5, 1148) \approx 3.7217$

$\frac{\partial f}{\partial h} = 0.00091s[0.103(\ln 2.505)(2.505^h)]$, $\frac{\partial f}{\partial s} = 0.00091[0.103(2.5^h) + 1]$

When $h = 3.5$, $s = 1148$, and $\Delta h = -0.5$, $\frac{ds}{dh} = \frac{-f_h}{f_s} \approx -758.2811$ and

$\Delta s \approx \frac{ds}{dh}\Delta h \approx (-758.2811)(-0.5) \approx 379.14$.

The input s should increase by approximately 379.14 in order to compensate for a decrease of 0.05 in h.

13. a. $A(6, 250) \approx 7.16$

The average cost is approximately \$7.16.

b. $\frac{\partial A}{\partial n} = (-0.02c^2 + 0.35c + 0.99)(\ln 0.99897)(0.99897^n)$

When $c = 6$ and $n = 250$,

$\frac{\partial A}{\partial n} = (-0.02c^2 + 0.35c + 0.99)(\ln 0.99897)(0.99897^n) \approx -0.001888$

the average cost is changing at a rate of approximately –\$0.002 per shirt.

c. $\frac{\partial A}{\partial c} = (-0.04c + 0.35)(0.99897^n) + 0.46$

$\frac{dn}{dc} = \frac{-A_c}{A_n} = \frac{-[(-0.04c + 0.35)(0.99897^n) + 0.46]}{(-0.02c^2 + 0.35c + 0.99)(\ln 0.99897)(0.99897^n)}$ shirts per color

We expect $\frac{dn}{dc}$ to be positive because if the number of colors increases, the order size would also need to increase to keep average cost constant.

d. When $c = 4$ and $n = 500$, $\frac{dn}{dc} \approx 450$ shirts per color. For each additional color, the order size would need to increase by approximately 450 shirts. Similarly, if the number of colors decreases by 1, the order size could decrease by approximately 450 shirts and the average cost would remain constant.

15. a, b.

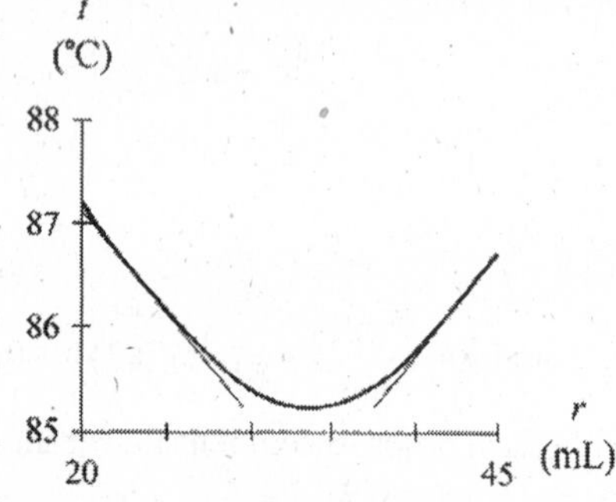

c. $\frac{\partial p}{\partial t} = -9.6544 + 0.14736t$, $\frac{\partial p}{\partial r} = 1.9836 - 0.05916r$

$\frac{dt}{dr} = \frac{-p_r}{p_t} = \frac{-(1.9836 - 0.05916r)}{-9.6544 + 0.14736t}$ °C per milliliter

d. We can solve for r in $p(86.5, r) = 53$ to get $r \approx 23.125$ or $r \approx 43.934$.

When $t = 86.5$ and $r = 23.125$, $\frac{dt}{dr} \approx -0.199$ °C per milliliter

When $t = 86.5$ and $r = 43.934$, $\frac{dt}{dr} \approx 0.199$ °C per milliliter

It is also possible to calculate the slope formula as

$\frac{dr}{dt} = \frac{-p_t}{p_r} = \frac{-(-9.6544 + 0.14736t)}{1.9836 - 0.05916r}$ milliliter per °C and the two specific slopes as follows:

When $t = 86.5$ and $r = 23.125$, $\frac{dr}{dt} \approx -5$ milliliter per °C

When $t = 86.5$ and $r = 43.934$, $\frac{dt}{dr} \approx 5$ milliliter per °C

To illustrate the tangent lines whose slopes these values represent, we must draw the parabola sideways with one tangent line on top and one on the bottom.

17. a. $B(67, 129) \approx 20.044$ points

Solving $B(h, w) = 20.044$ for w, we get $w = \frac{(20.044)(0.00064516)h^2}{0.45}$ pounds

b. $\frac{dw}{dh} = \frac{2(20.044)(0.00064516)h}{0.45}$

When $h = 67$, $\frac{dw}{dh} = \frac{2(20.044)(0.00064516)(67)}{0.45} \approx 3.85$ pounds per inch

c. The answer to part *b* agrees with the answer given in Example 1.

19. a. $m(10{,}000, 5) \approx \$193.31$

b. $\frac{\partial m}{\partial A} = \frac{0.005}{1-0.9419^t}$, $\frac{\partial m}{\partial t} = \frac{0.005A}{(1-0.9419^t)^2}(\ln 0.9419)0.9419^t$

When $A = 10{,}000$, $t = 5$, and $\Delta t = -1$, $\Delta A = \frac{-m_t}{m_A}\Delta t \approx -\$2212.65$

The amount you could borrow is approximately \$10,000 – \$2212.65 = \$7787.35.

21. a. From the table, we estimate that Coke would need to lower its prices by more than \$0.50 per can.

b. Using the equations $S(1.00, P) = 196P + 25$ cans of Coke products and $S(c, 1.25) = -50.286C^2 + 7.771C + 312.6$ cans of Coke products where $\$P$ is the price of Pepsi products and $\$C$ is the price of Coke products, we estimate the change in Coke prices as $\Delta C \approx \frac{-S_P}{S_C}\Delta P = \left(\frac{-196}{-92.8}\right)(-0.25) \approx -\$0.53$. Coke would need to lower its price from \$1.00 a can to approximately \$0.47 a can.

c. Rather than lower its prices so drastically, Coke should probably consider such alternatives as more advertising on campus.

23. $\frac{f_x}{f_y}$ represents the opposite of the slopes of contour curves along the contour graph of $f(x, y)$ where y is graphed along the vertical axis and x is graphed along the horizontal axis.

Chapter 9 Concept Review

1. a.

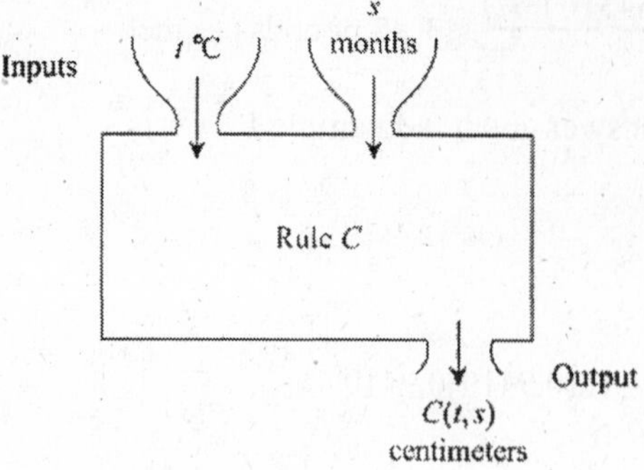

b. $C(3, 71) = 2.9$ cm
When apples are stored for 3 months and blanched at 71°C, the applesauce flows 2.9 cm down a vertical surface in 30 seconds.

c. Find a model for the consistometer value as function of storage time, using storage time as input and the values in the 35°C column as output. A piecewise continuous model is

$$C(s,35) = \begin{cases} -0.05s^2 + 0.35s + 3 \text{ cm} & \text{when } 0 \le s < 2 \\ -0.05s^2 + 0.15s + 3.4 \text{ cm} & \text{when } 2 \le s \le 4 \end{cases}$$

Assuming that 1 month is 4.3 weeks, 2 weeks corresponds to $s \approx \dfrac{2 \text{ wks}}{4.3 \text{ wks/month}}$
≈ 0.465 month

$C(0.465, 35) \approx 3.2$ centimeters

d. $\dfrac{\partial C(4,t)}{\partial t}$ is the rate of change of the consistometer value with respect to the blanching temperature when the storage time is 4 months.

e. Find a model for the consistometer value as function of the blanching temperature, using temperature as the input and the values in the row corresponding to a 4-month storage time as the output. A quadratic model is $C(4,t) = (6.9444 \cdot 10^{-4})t^2 - 0.0869t + 5.4124$ centimeters for $35 \le t \le 83$ where t is the blanching temperature in degrees Celsius.

$$\frac{\partial C(4,t)}{\partial t} = 2(6.9444 \cdot 10^{-4})t - 0.0869 \text{ cm per °C}$$

$$\frac{\partial C(4,45)}{\partial t} = 2(6.9444 \cdot 10^{-4})(45) - 0.0869 \approx -0.0244 \text{ cm per °C}$$

When the storage time is a constant 4 months and the blanching temperature is 45°C, the consistometer value is decreasing by approximately 0.024 cm per °C. That is, if the blanching temperature is increased to 46°C, the consistometer value should decrease by approximately 0.024 cm.

f.

Storage time (months)	Temperature (Celsius)				
	35°	47°	59°	71°	83°
0	3.0	2.8	2.6	2.6	2.8
1	3.3	3.1	2.8	2.8	3.0
2	3.5	3.2	3.0	2.9	3.2
3	3.4	3.2	3.0	2.9	3.2
4	3.2	2.9	2.7	2.7	3.0

3.0 3.2 2.8 3.0 3.2 2.8

2. a. $E(24, 60) \approx 466$ eggs
A female insect kept at 24°C and 60% relative humidity will lay approximately 466 eggs in 30 days.

b. $\frac{\partial E}{\partial h} = 23.1412 - 0.1874h - 0.4023t$

When $t = 27$°C and $h = 77\%$, $\frac{\partial E}{\partial h} = 23.1412 - 0.1874(77) - 0.4023(27) \approx 2.2$ eggs per percentage point of humidity. When the temperature is held constant at 27°C and the relative humidity is 77%, the number of eggs is decreasing by approximately 2.2 eggs per percentage of relative humidity. That is, if the relative humidity were increased to 78%, the number of eggs would decrease by approximately 2.

c. $\frac{\partial E}{\partial t} = 299.7038 - 10.4420t - 0.4023h$

$$\frac{dh}{dt} = \frac{-E_t}{E_h} = \frac{-(299.7038 - 10.4420t - 0.4023h)}{23.1412 - 0.1874h - 0.4023t} \text{ percentage points per °C}$$

d. When $h = 63\%$ and $t = 25$°C, $\frac{dh}{dt} \approx -10.4$ percentage points per °C.

$\Delta h \approx \frac{dh}{dt}\Delta t \approx (-10.4)(-0.5) = 5.2$ percentage points
The humidity should increase by approximately 5.2%.

e. On the contour curve corresponding to the egg production for t =25°C and h = 63% (the 490.1601 egg contour curve), the slope of the tangent line at that point is −10.4 percentage points per °C. The h value on the tangent line when t = 24.5°C is $h \approx 63 + 5.2 = 68.2\%$. This is an approximation to the value of h that corresponds to t = 24.5°C on the 490.1601 egg contour curve.

3. a. Inputs are $t \approx 26$ °C and $h \approx 56\%$. The output is approximately 485 eggs.

b. An increase in temperature by 3°C will result in a greater change in the number of eggs laid.

c. The number of eggs laid will decrease more rapidly when the temperature decreases than when the humidity decreases.

d. By sketching a tangent line on Figure 9.42 and calculating its slope, you should obtain an estimate of approximately 13 percentage points per °C.

e. $\dfrac{dh}{dt} = \dfrac{-E_t}{E_h} = \dfrac{-24.1532}{1.8672} \approx -12.9$ percentage points per °C

When the temperature is 24°C and the humidity is 62%, the change in humidity needed to compensate for a small change in temperature (so that the number of eggs remains constant) can be estimated as $\Delta h \approx (-12.9)\Delta t$ percentage points.

4. From Activity 2 we have $\dfrac{\partial E}{\partial h} = 23.1412 - 0.1874h - 0.4023t$

$$\frac{\partial E}{\partial t} = 299.7038 - 10.4420t - 0.4023h$$

$E_{tt} = -10.4420$, $E_{th} = -0.4023$, $E_{hh} = -0.1874$, and $E_{ht} = -0.4023$

The second partials matrix for any values of h and t is $\begin{array}{c} \\ t \\ h \end{array}\begin{array}{c} \begin{array}{cc} t & h \end{array} \\ \begin{bmatrix} -10.442 & -0.4023 \\ -0.4023 & -0.1874 \end{bmatrix} \end{array}$.

Chapter 10
Analyzing Multivariable Change: Optimization

Section 10.1 Multivariable Critical Points

1. **a.** A relative maximum occurs when a table value is greater than all 8 values surrounding it.

   b. A relative minimum occurs when a table value is less than all 8 values surrounding it.

   c. If a table value appears to be a maximum in one direction but a minimum in another direction, then the value corresponds to a saddle point.

   d, e. If all the edges of a table are terminal edges, then the absolute maximum and minimum are simply the largest and smallest values in the table. If all the edges are not terminal edges, then you must know whether any critical points exist outside the table in order to determine whether absolute extrema exist. If no critical points exist outside the table, then in determining absolute extrema, you must consider relative extrema, output values on terminal edges, and the behavior of the function beyond the edges of the table.
   It is often helpful to sketch contour curves on a table when determining critical points and absolute extrema.

3. The point is a relative maximum point because the values of the contour curves decrease in all directions away from the point.

5. **a.**

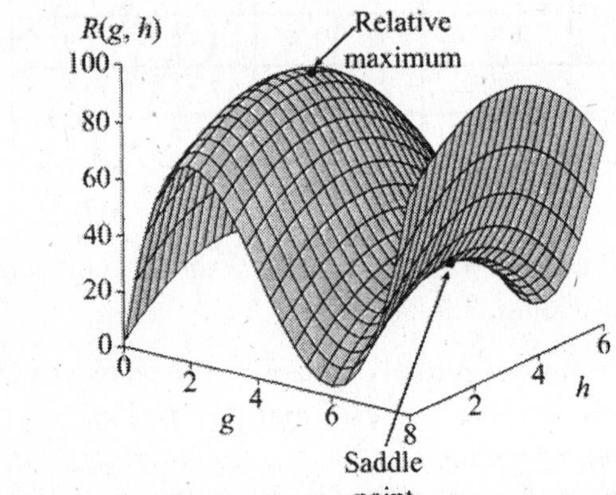

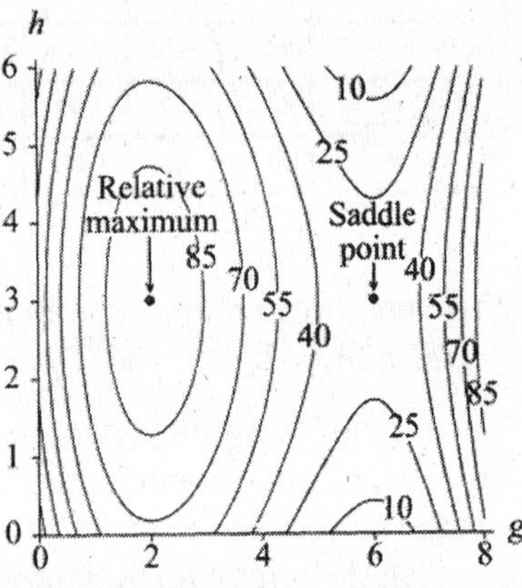

   b. Relative maximum point: $(g, h, R) \approx (2, 3, 95)$; Saddle point: $(g, h, R) \approx (6, 3, 30)$

7. The point is a saddle point because it is a maximum contour level along a cross-section extending from the x=4, y=0 corner diagonally back through the point and is a minimum contour level along a cross-section extending from the x=0, y=0 corner diagonally through the point.

9. Relative maximum point: (May, 1995, $1.45 per pound)
   Relative maximum point: (May, 1998, $0.88 per pound)

11. **a.** Yes; The table gives monthly averages, so it doesn't make sense to extend the columns. However, the choice of January as the first column is not mandatory. The best way to visualize this table is to wrap it around a cylinder so that the January and December columns are adjacent columns and there are no left or right edges on the table. The top and bottom rows are terminal edges.

b.

	Jan	Feb	Mar	Apr	May	Jun	Jul	Aug	Sep	Oct	Nov	Dec
North Pole	—	—	—	3.1	6.9	8.9	7.9	4.9	0.80	—	—	—
80°	—	—	0.77	3.4	6.9	8.8	7.8	4.8	1.55	0.13	—	—
70°	—	0.51	2.0	4.5	7.[illegible]	8.5	7.8	5.5	2.9	0.95	0.06	—
60°	0.55	1.53	3.4	5.7	7.7	8.8	8.2	6.5	4.3	2.2	0.81	0.34
50°	1.66	2.8	4.7	6.7	8.4	9.1	8.8	7.4	5.5	3.4	1.97	1.35
40°	3.0	4.2	5.9	7.5	8.8	9.3	9.0	8.1	6.5	4.8	3.4	2.6
30°	4.4	5.6	6.9	8.1	9.0	9.2	9.1	8.4	7.4	6.1	4.7	4.1
20°	5.8	6.7	7.8	8.5	8.8	8.9	8.8	8.6	8.0	7.1	6.0	5.5
10°	7.1	7.7	8.3	8.5	8.4	8.3	8.3	8.4	8.3	7.9	7.2	6.8
Equator	8.1	8.5	8.6	8.3	7.8	7.5	7.6	8.0	8.4	8.4	8.2	7.9
10°	8.9	8.8	8.4	7.7	6.9	6.4	6.5	7.2	8.1	8.6	8.8	8.8
20°	9.4	9.0	8.1	6.9	5.7	5.1	5.4	6.3	7.5	8.6	9.2	9.5
30°	9.6	8.8	7.4	5.8	4.4	3.8	4.1	5.2	6.7	8.2	9.3	9.8
40°	9.6	8.3	6.5	4.6	3.1	2.5	2.7	3.9	5.6	7.5	9.1	9.9
50°	9.3	7.6	5.4	3.3	1.84	1.25	1.49	2.6	4.5	6.6	8.7	9.7
60°	8.7	6.8	4.1	2.0	0.72	0.31	0.47	1.36	3.1	5.6	8.0	9.3
70°	8.2	5.5	2.8	0.84	—	—	—	0.38	1.78	4.3	7.2	9.1
80°	8.2	4.7	1.42	0.099	—	—	—	—	0.62	3.2	7.0	9.3
South Pole	8.1	4.6	0.60	—	—	—	—	—	—	2.9	7.[illegible]	9.4

c. There are three relative maximum points: (June, North Pole, 8.9 kW-h/m^2), (June, 40° North, 9.3 kW-h/m^2), and (December, 40° South, 9.9 kW-h/m^2).

It is difficult to estimate relative minima points accurately because of the dashes in the table, which we can interpret to mean radiation levels of essentially zero. Thus we conclude that the regions of the underlying function represented by the dashes in the table are those in which the minimum radiation level occurs. There are two such regions: one at and near the North Pole between March and October and one at and near the South Pole between April and September. If there are two specific relative minima of the underlying function, then we estimate that they occur at the end of December at the North Pole and in the middle of June at the South Pole.

There are four points that can be considered saddle points: (April, 10° North, 8.5 kW-h/m^2) (August, 10° North, 8.4 kW-h/m^2), (June, 70° North, 8.5 kW-h/m^2), and (December, 70° South, 9.1 kW-h/m^2). (Answers may vary.)

d. The greatest radiation level shown in the table is 9.9 kW-h/m^2 which occurs in December at 40° South. The smallest radiation level shown is 0.06 kW-h/m^2 which occurs in November at 80° North. If we consider the dashes to be zeros, then the smallest radiation level is zero and occurs many times in the table.

e. The absolute maximum value is 9.9 kW-h/m^2, and the absolute minimum is approximately zero. Because the table cannot extend in any direction, these answers do correspond to those in part *d.*

f. The largest and smallest values in the table will be the absolute maximum and minimum, respectively, if the table cannot be extended in any direction. That is, either the edges are terminal edges or the table "wraps around," as in this case.

13. a. The expected corn yield is 100% of the annual average yield. That is, there is no expected increase or decrease in yield from the average.

b. The expected corn yield is 40% of the annual average yield.

c.

T (°C) \ *P* (%)	-100	-90	-80	-70	-60	-50	-40	-30	-20	-10	0	10	20	30	40	50	60	70	80
6	0	7	14	18	23	27	32	40	48	56	64	67	69	72	74	75	75	75	76
	0	7	15	20	25	29	34	43	51	59	67	70	73	75	78	78	78	78	78
5	0	8	16	21	26	32	37	45	54	62	71	73	76	79	81	81	81	81	81
	0	8	17	22	28	34	39	48	57	65	74	77	79	82	85	85	84	84	84
4	0	9	18	24	30	36	42	51	60	68	77	80	83	86	88	88	88	87	87
	0	9	19	25	32	38	45	54	62	71	80	83	86	89	92	91	91	90	89
3	0	10	20	27	33	40	47	56	65	75	84	87	90	93	96	95	94	93	92
	0	10	21	28	35	42	49	59	69	78	87	91	94	96	98	97	95	94	93
2	0	11	21	29	37	44	52	63	74	82	90	95	99	99	100	98	97	96	94
	0	11	22	30	38	46	54	65	76	85	93	99	101	101	102	100	99	97	95
1	0	12	23	31	40	48	56	67	78	87	95	102	103	104	104	102	100	98	96
	0	12	24	33	41	50	58	69	80	90	98	104	105	106	106	104	102	100	97
0	0	12	25	34	43	51	60	71	83	92	100	107	107	108	109	106	103	101	98
	0	13	26	35	43	52	61	72	84	93	101	107	108	108	107	105	102	100	97
-1	0	13	27	35	44	53	61	73	85	94	103	108	109	108	106	104	101	99	96
	0	14	28	36	45	53	62	73	84	93	102	107	108	107	105	103	100	98	95
-2	0	14	29	37	46	54	63	73	84	93	101	105	107	106	104	102	99	97	94
	0	15	29	38	46	55	63	72	82	90	98	101	103	103	103	101	98	95	93
-3	0	15	30	39	47	55	64	72	80	88	96	97	99	101	102	100	97	94	92
	0	15	30	38	46	54	62	70	77	85	92	94	95	98	98	96	94	91	89
-4	0	15	31	38	46	53	61	68	74	81	88	90	91	93	95	92	90	88	85
	0	15	31	38	45	52	59	65	72	78	85	86	88	89	91	89	87	85	82
-5	0	15	31	38	44	51	58	63	69	75	81	82	84	85	87	85	83	81	79
	0	16	31	37	44	50	56	61	67	72	77	79	80	82	83	81	80	78	76
-6	0	16	31	37	43	49	55	59	64	69	74	75	76	78	79	78	76	75	73

d. The maximum percentage yield is 109%. This maximum occurs twice, at the points (40%, 0°C, 109%) and (20%, –1°C, 109%). This means that a yield of 109% above normal can be expected when temperatures are average (a change of 0°C) and there is 40% more precipitation than normal or when temperatures are 1°C below normal and precipitation is 20% above normal.

15. a. Because there are no values smaller than all 8 surrounding values in the table, there are no relative minimum points. A relative maximum point occurs at an average daily weight gain of 1.01 kilograms per day for a 91-kilogram pig at an air temperature of 21.1°C.

When we consider the edges of the table in our search for absolute extrema, we find the absolute maximum weight gain to be 1.09 kilograms per day for a 156-kilogram pig and air temperature of 15.6°C. The absolute minimum point corresponds to a weight loss of 1.15 kilograms per day for a 156-kilogram pig and temperature of 37.8°C.

b. 15.6°C = 60.08°F, 21.1°C = 69.98°F, 37.8°C = 100.04°F

c. The greatest average daily weight gain for pigs weighing between 45 kg and 156 kg and air temperatures between 4.4°C and 37.8°C is approximately 1.09 kilograms per day for a 156-

kilogram pig at an air temperature of about 60°F. The greatest average daily weight loss is 1.15 kilograms per day for a 156-kilogram pig at an air temperature of about 100°C. These answers indicate that the heaviest pigs can gain or lose weight more quickly than lighter pigs, depending on the temperature.

17. a.

Storage time (months)	Temperature (°C)				
	35	47	59	71	83
0	3.0	2.8	2.6	2.6	2.8
1	3.3	3.1	2.8	2.8	3.0
2	3.5	3.2	3.0	2.9	3.2
3	3.4	3.2	3.0	2.9	3.2
4	3.2	2.9	2.7	2.7	3.0

b. Saddle point:

≈(71 °C, 2.5 months, 2.9 cm)

c. Absolute maximum: 3.5 cm at 35°C and 2 months

Absolute minimum: 2.6 cm at 59°C and 71°C and 0 months

d. *One possible answer:* The saddle point is important because it shows the optimal storage time and temperature to keep applesauce from getting too thick or too thin.

19. a.

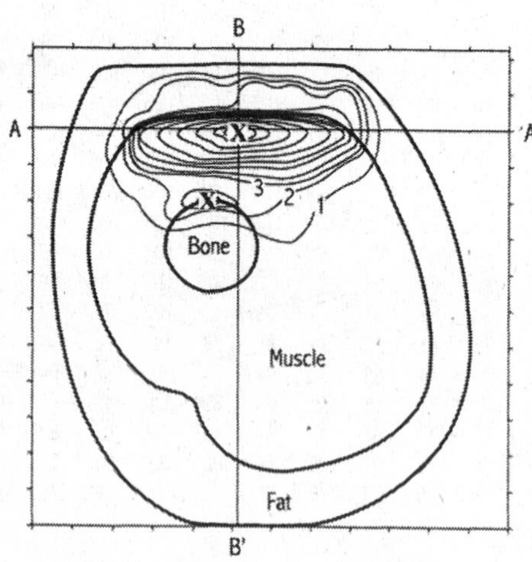

b. Both points correspond to maximum temperatures.

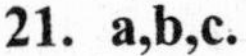

21. a,b,c.

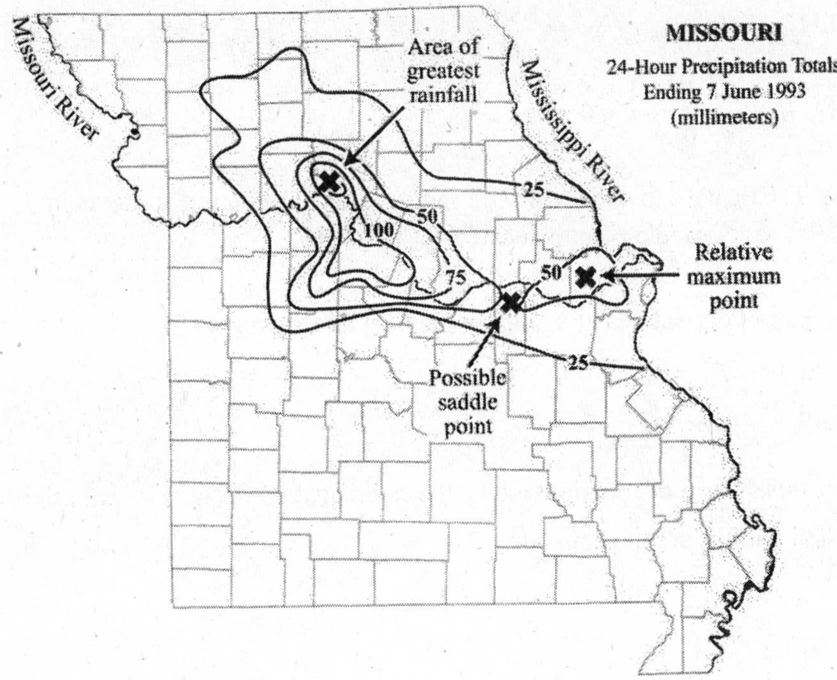

Section 10.2 Multivariable Optimization

1. Find the partial derivatives of R, and set them equal to zero.

$R_k = 6k - 2m - 20 = 0$

$R_m = -2k + 6m - 4 = 0$

Solving for k and m gives $k = 4$ and $m = 2$.

$R_{kk} = 6$, $R_{mm} = 6$, $R_{km} = R_{mk} = -2$

$$D(4,\ 2) = \begin{vmatrix} 6 & -2 \\ -2 & 6 \end{vmatrix} = 36 - 4 = 32 > 0$$

$R_{kk}(4,\ 2) = 6 > 0$

Because $D > 0$ and $R_{kk} > 0$, we know the critical point corresponds to a minimum. We conclude that a relative minimum of $R(4, 2) = 16$ is located at $k = 4$ and $m = 2$.

3. Find the partial derivatives of G, and set them equal to zero.

$G_t = pe^t = 0$

$G_p = e^t - 3 = 0$

To solve for t and p, note that in order for $pe^t = 0$, either $p = 0$ or $e^t = 0$. Because e^t can never be zero, we conclude that $p = 0$. The second equation gives $e^t = 3$ or $t = \ln 3 \approx 1.099$.

$G_{tt} = pe^t$, $G_{pp} = 0$, $G_{tp} = G_{pt} = e^t$

At $p = 0$ and $t = \ln 3$,

$G_{tt} = 0$, $G_{pp} = 0$, $G_{tp} = G_{pt} = e^{\ln 3} = 3$

$$D(\ln 3,\ 0) = \begin{vmatrix} 0 & 3 \\ 3 & 0 \end{vmatrix} = -9 < 0$$

Because $D < 0$, we know the point is a saddle point. The output at that point is $G(\ln 3, 0) = 0$. A saddle point is located at (ln 3, 0, 0).

5. Find the partial derivatives of h, and set them equal to zero.

$h_w = 1.2w - 4.7z = 0$

$h_z = 3.0z^2 - 4.7w = 0$

To solve this system of equations, solve the first equation for w: $w = \frac{4.7z}{1.2}$. Substitute this expression into the second equation:

$$3.0z^2 - 4.7\left(\frac{4.7z}{1.2}\right) = 0$$

$$z\left(3.0z - \frac{4.7^2}{1.2}\right) = 0$$

Solving for z gives two solutions: $z = 0$ and $z \approx 4.720$. The two critical points are
$w = 0$, $z = 0$, $h = 0$ and
$w \approx 18.487$, $z \approx 4.720$, and $h \approx -68.353$

The second partials are

$h_{ww} = 1.2$, $h_{zz} = 6.0z$

$h_{zw} = h_{wz} = -4.7$

$D(0,\ 0) = \begin{vmatrix} 1.2 & -4.7 \\ -4.7 & 0 \end{vmatrix} = -20.89 < 0$, thus (0, 0, 0) is a saddle point.

$D(18.487,\ 4.720) = \begin{vmatrix} 1.2 & -4.7 \\ -4.7 & 28.32 \end{vmatrix} \approx 11.89 > 0$ and $h_{ww} > 0$, thus (18.487, 4.720, –68.353) is a relative minimum.

7. Solving $f_x = 6x - 3x^2 = 3x(2 - x) = 0$ and $f_y = 24y - 24y^2 = 24y(1 - y) = 0$ yields $x = 0$, $x = 2$, $y = 0$, $y = 1$. Thus we consider 4 points: (0, 0), (0, 1), (2, 0), (2, 1). The second partials are $f_{xx} = 6 - 6x$, $f_{yy} = 24 - 48y$, and $f_{xy} = f_{yx} = 0$.

For (0, 0), $D = \begin{vmatrix} 6 & 0 \\ 0 & 24 \end{vmatrix} = 144 > 0$ and $f_{xx} > 0$ thus (0, 0, 60) is a relative minimum.

For (0, 1), $D = \begin{vmatrix} 6 & 0 \\ 0 & -24 \end{vmatrix} = -144 < 0$, thus (0, 1, 64) is a saddle point.

For (2, 0), $D = \begin{vmatrix} -6 & 0 \\ 0 & 24 \end{vmatrix} = -144 < 0$,
thus (2, 0, 64) is a saddle point.

For (2, 1), $D = \begin{vmatrix} -6 & 0 \\ 0 & -24 \end{vmatrix} = 144 > 0$ and $f_{xx} < 0$, thus (2, 1, 68) is a relative maximum.

9. a. Solving $R_b = 14 - 6b - p = 0$ and $R_p = -b - 4p + 12 = 0$ yields
$b \approx 1.91$ and $p \approx 2.52$.
The manager should try to buy ground beef at \$1.91 a pound and sausage at \$2.52 a pound.

b. We verify that these inputs give a maximum revenue by finding the determinant of the second partials matrix: $R_{bb} = -6$, $R_{pp} = -4$, $R_{bp} = R_{pb} = -1$

$$D = \begin{vmatrix} -6 & -1 \\ -1 & -4 \end{vmatrix} = 23 > 0 \text{ and } R_{bb} < 0$$

Thus we have found the prices that result in maximum quarterly revenue.

c. $R(1.91,\ 2.52) \approx \$28.5$ thousand

11. a. Find the partial derivatives of E, and set them equal to zero.

$$E_T = 299.7038 - 10.4420T - 0.4023H = 0$$
$$E_H = 23.1412 - 0.1874H - 0.4023T = 0$$

Solving for T and H gives
$T \approx 26.1032°\text{C}$ and $H \approx 67.4488\%$.
$E(26.1032, 67.4488) \approx 500.343$ eggs The critical point is approximately (26.1, 67.4, 500).

b. When exposed to approximately 26.1 °C and 67.4% relative humidity, a *C. grandis* female will lay approximately 500 eggs in 30 days.
$E_{TT} = -10.442$, $E_{RR} = -1.874$, $E_{TR} = E_{RT} = -0.4023$

$$D = \begin{vmatrix} -10.442 & -0.4023 \\ -0.4023 & -1.874 \end{vmatrix} \approx 19.4 > 0$$

Because $E_{TT} < 0$ and $D > 0$, the critical point is a maximum.

13. a. Find the partial derivatives of R, and set them equal to zero.

$$R_P = -1.544 + 9.810P - 3T = 0$$
$$R_T = -1.625 - 14.106T - 3P = 0$$

Solving for P and T gives $P \approx -0.1307$ and $T \approx -0.0874$.
$R(-0.1307, -0.0874) \approx 32.8$
The critical point is approximately (−0.13, −0.09, 32.8).

b. The second partials are
$R_{PP} = -9.810$, $R_{TT} = -14.106$, $R_{PT} = R_{TP} = -3$

$$D = \begin{vmatrix} -9.810 & -3 \\ -3 & -14.106 \end{vmatrix} \approx 129 > 0$$

Because $D > 0$ and $R_{PP} < 0$, the critical point is a maximum.

To maximize the rate, the pH is about $5.5 + 1.5(-0.13) \approx 5.3$ and the temperature is about $60 + 8(-0.09) \approx 59.3°\text{C}$.

15. a. From the graph, the maximum appears to be about 2.35 mg when pH is about 9 and the temperature is about 65 °C.

b. Find the partial derivatives of P, and set them equal to zero.

$P_x = -0.26 - 0.46x - 0.25y = 0$

$P_y = -0.34 - 0.32y - 0.25x = 0$

Solving for x and y gives $x \approx 0.0213$ and $y \approx -1.0791$.
The second partials are $P_{xx} = -0.46$, $P_{yy} = -0.32$, and $P_{xy} = P_{yx} = -0.25$

$$D = \begin{vmatrix} -0.46 & -0.25 \\ -0.25 & -0.32 \end{vmatrix} \approx 0.21 > 0$$

Because $P_{xx} < 0$ and $D > 0$, the critical point is a maximum.

A pH of about $9 + 0.0213 \approx 9.02$ and a temperature of about $70 + 5(-1.0791) \approx 64.6$°F will result in the maximum production of $P(0.0213, -1.0791) \approx 2.32$ mg.

17. Find the partial derivatives of L, and set them equal to zero.

$L_w = 1.13 - 5.83s = 0$

$L_s = 1.04 - 5.83w = 0$

Solving for w and s gives $w \approx 0.18$ and $s \approx 0.19$. The second partials are

$L_{ww} = 0$, $L_{ss} = 0$, $L_{sw} = L_{ws} = -5.83$

$$D = \begin{vmatrix} 0 & -5.83 \\ -5.83 & 0 \end{vmatrix} \approx -34 < 0$$

Because $D < 0$, the critical point is a saddle point. The corresponding proportions are
whey protein: $w \approx 0.18$
skim milk powder: $s \approx 0.19$
sodium caseinate: $c \approx 1 - 0.18 - 0.19 = 0.63$

19. a. Find the partial derivatives of E, and set them equal to zero.

$E_e = -30.372e^2 + 42.694e - 13.972 = 0$

$E_n = -5n + 2.497 = 0$

Solving for n gives $n \approx 0.4994$ and solving for e gives $e \approx 0.5185$ or $e \approx 0.8872$. Thus there are two critical points:

$A \approx (0.5185, 0.4994, 799.91)$
$B \approx (0.8872, 0.4994, 800.16)$.

b. We use the contour graph to conclude that Point A is a saddle point and point B corresponds to a relative maximum.

c. $E_{ee} = -60.744e + 42.694$, $E_{nn} = -5$, $E_{ne} = E_{en} = 0$
When $e \approx 0.5185$ and $n \approx 0.4994$,
$D \approx -59.98$ so point A is confirmed to be a saddle point. When $e \approx 0.8872$ and $n \approx 0.4994$,
$D \approx 59.98$ and $e \approx -11.20$ so point B is confirmed to correspond to relative maximum.

21. a. Find the partial derivatives of A, and set them equal to zero.

$A_g = 4.26 - 0.1g = 0$

$A_m = 5.69 - 0.28m - 0.07h = 0$

$A_s = 0.67 - 0.06s = 0$

$A_h = 2.48 - 0.1h - 0.07m = 0$

Solving this system, we get

$g = 42.6\%$, $m \approx 17.1169\%$,

$s \approx 11.1667\%$, $h \approx$ 12.8182 days

$A(42.6, 7.1169, 11.1667, 12.8182) \approx 7.29$

b. One method is to evaluate points close to the critical point. By doing so, it is possible to conjecture that the point is a relative maximum.

23. *One possible answer:* In order to locate a critical point of a three-dimensional function, determine both first partial derivatives, set each of them equal to zero and solve. Any input (a, b) for which both first partial derivatives are zero will yield a critical point.

To determine the type of critical point found, write the four second partial derivatives of the function. Next evaluate the product of the non-mixed second partials minus the product of the mixed partials at (a, b)—this is known as the $D(a, b)$, the determinant of the second partials matrix. If $D(a, b)$ is negative, then (a, b) yields a saddle point. If $D(a, b)$ is positive, then evaluate one of the non-mixed second partials at (a, b). If that second partial is negative, then (a, b) yields a maximum. If it is positive, then (a, b) yields a minimum. Finally, if $D(a, b)$ is exactly zero, then the determinant test fails and estimation graphically or numerically may help.

Section 10.3 Optimization Under Constraints

1. a. The optimal point is (45, 45, 2025). The optimal value is 2025.

b. The point is a constrained maximum.

c. $f_a = b$, $f_b = a$, $g_a = 1$, $g_b = 1$

We have the following system of equations:

$$b = \lambda(1)$$
$$a = \lambda(1)$$
$$a + b = 90$$

Solving this system, we get $a = 45$, $b = 45$, and $\lambda = 45$. Thus $f(45, 45) = 2025$ is a constrained optimal value.

3. a.

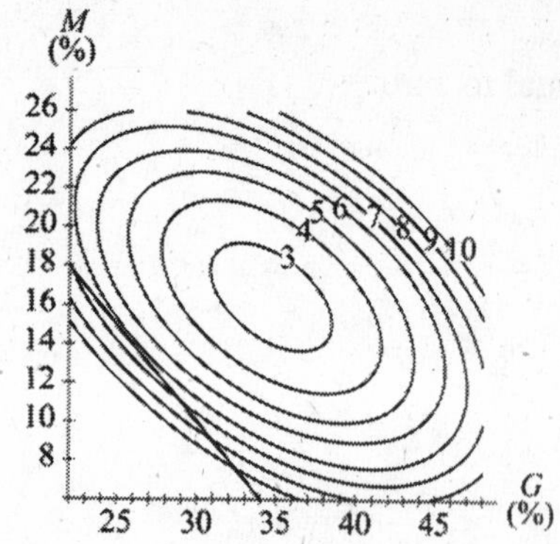

The constrained minimum is approximately 7, which occurs when $G \approx 26\%$ and $M \approx 15\%$.

5. $f_r = 4r + p$, $f_h = r - 2p + 1$,
$g_k = 2$, $g_h = 3$
We have the following system of equations.

$$4r + p = \lambda(2)$$
$$r - 2p + 1 = \lambda(3)$$
$$2r + 3p = 1$$

Solving this system we get $r = \frac{-1}{16}$ and $p = \frac{3}{8}$. $f\left(\frac{-1}{16}, \frac{3}{8}\right) = \frac{7}{32}$

The constrained optimal point is $\left(\frac{-1}{16}, \frac{3}{8}, \frac{7}{32}\right)$.

a.

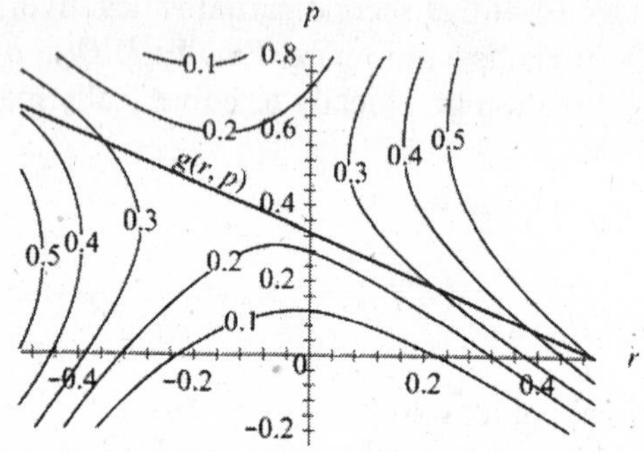

A contour graph confirms that the point is a constrained minimum, because the contour to which the constraint line is tangent is the smallest-valued contour that the constraint line touches.

b. We evaluate $f(r, p)$ for values of r near $\frac{-1}{16} = 0.0625$. We choose 0.6 and 0.65 (these are arbitrarily chosen). We find the corresponding p-values that lie on the constraint curve $2r + 3p = 1$, but substituting each r-value and solving for p. Thus we have the input pairs (0.065,0.29) and (0.06, 0.2933). Evaluating f at these inputs gives the outputs 0.2321 and 0.2332. Both of these values are greater than $f\left(\frac{-1}{16}, \frac{3}{8}\right) = \frac{7}{32} = 0.21875$. These calculations suggest that the point is the location of a constrained minimum.

7. We have the following system of equations (the first two are the same as in Example 2).

$$1.13 - 5.83s = \lambda$$
$$1.04 - 5.83w = \lambda$$
$$w + s = 0.9$$

Solving this system gives $s \approx 0.4577$ and $w \approx 0.4423$. These values correspond to an output $P(0.4577, 0.4423) \approx 10.45$, so the minimum percentage loss is about 10.45%. The approximation in Example 2 was 10.46%.

9. **a.** $g(r, n) = 12r + 6n = 504$

$A_r = 0.2rn$, $A_n = 0.1r^2$, $g_r = 12$, $g_n = 6$

We have the following system of equations.

$$0.2rn = \lambda(12)$$
$$0.1r^2 = \lambda(6)$$
$$12r + 6n = 504$$

Solving this system, we get $r = 28$, $n = 28$, and $\lambda \approx 13.07$ (or $r = 0$, $n = 0$, and $\lambda = 0$ which gives 0 responses.) The club should allocate (28 ads)($12 per ad) = $336 for 28 radio ads and (28 ads)($6 per ad) = $168 for 28 newspaper ads.

b. $A(28, 28) \approx 2195$ responses

c. The Lagrange multiplier is $\lambda = 13.07$ responses per dollar. The change in the number of responses can be approximated as

$\Delta A \approx$ (13.07 responses per dollar)($26) $\approx$ 340 additional responses.

11. **a.** We solve the system of equations: $C_G = -3.76 + 0.08G + 0.06M = 0$ $C_M = -4.71 + 0.16M + 0.06G = 0$ to obtain $G = 34.7\%$, $M \approx 16.4\%$, and an absolute minimum cohesiveness of honey of $C(34.7, 16.4) \approx 2.5$. Our estimate in Activity 3 part *a* was 2.5, agreeing to one decimal place with the actual minimum.

b. $g(G, M) = G + M = 40$

Using the partial derivatives from part *a* and $g_G = 1$ and $g_M = 1$, we have the following system of equations:

$$-3.76 + 0.08G + 0.06M = \lambda(1)$$
$$-4.71 + 0.16M + 0.06G = \lambda(1)$$
$$G + M = 40$$

Solving this system, we get $G \approx 25.4$, $M \approx 14.6$, and $\lambda \approx -0.85$.

The minimum measure of cohesiveness possible is about $C(25.4, 14.6) \approx 7.2$.

c. Figure 10.3.3 verifies that the point is a minimum since the contour to which the constraint line is tangent is the smallest-valued contour that the constraint line touches.

d. The estimate in Activity 3 part *b* of 7.5 is slightly higher than 7.2, the actual constrained minimum found in part *b* of this activity.

13. a. Worker expenditure:

$$\frac{(\$7.50/\text{hour})(100L \text{ hours})}{1000}$$

$= 0.75L$ thousand dollars

The constraint is

$g(L, K) = 0.75L + K$

The partial derivatives are

$f_L = 3.16389L^{-0.7}K^{0.5}$ $f_K = 5.27315L^{0.3}K^{-0.5}$

$g_L = 7.5$, $g_K = 1$

We solve the following system of equations

$$3.16389L^{-0.7}K^{0.5} = 0.75\lambda$$

$$5.27315L^{0.3}K^{-0.5} = \lambda$$

$$0.75L + K = 15$$

to obtain $L = 7.5$, $K = 9.375$, $\lambda \approx 3.152$.

Maximum production will be achieved by using 750 labor-hours and \$9375 in capital expenditures.

b. To verify that the value in part *a* is a maximum, evaluate the production at close points on the constraint curve, or examine the constraint curve graphed on a contour graph of the production function.

c. The marginal productivity of money is $\lambda \approx 3.152$ radios per thousand dollars. An increase in the budget of \$1000 will result in an increase in output of about 3 radios.

15. a. From Activity 11, $\lambda \approx -0.85$, so $\frac{dC}{dk} \approx -0.85$ cohesiveness unit per percentage point.

b. The minimum cohesiveness measure should decrease by about (0.85 unit per percentage point)(2 percentage points) = 1.7.

c. Find the partial derivatives of *C*, and set them equal to zero.

$C_G = -3.76 + 0.08G + 0.06M = 0$ $C_M = -4.71 + 0.16M + 0.06G = 0$

Solving for *G* and *M* gives

$G \approx 34.6739$ and $M \approx 16.4348$.

The second partials are

$C_{GG} = 0.08$, $C_{MM} = 0.16$, $C_{GM} = C_{MG} = 0.06$

$$D = \begin{vmatrix} 0.08 & 0.06 \\ 0.06 & 0.16 \end{vmatrix} = 0.0092 > 0$$

Because $C_{GG} > 0$ and $D > 0$, the critical point is a minimum.

$C(34.6739, 16.4348) \approx 3.08$

The relative minimum when there are no constraints is approximately 3.1 which is obtained when the percentage of glucose and maltose is approximately 34.7% and the percentage of moisture is approximately 16.4 %.

17. a. From Activity 13, $\lambda \approx 3.152$, so $\frac{dP}{dc} \approx 3.152$ radios/thousand dollars.

b. $\Delta P \approx$ (3.152 radios per thousand dollars)(1.5 thousand dollars)
≈ 4.7 radios

c. $\Delta P \approx$ (3.152 radios per thousand dollars)(–1 thousand dollars)
≈ -3.2 radios
$P(7.5, 9.375) \approx 30$ radios
We estimate the maximum production to be about $30 - 3 = 27$ radios.

19. a. $S(r,h) = 2\pi rh + \pi r^2 + \pi\left(r + \frac{9}{8}\right)^2$
square inches when the radius is r inches and the height is h inches

b. $V(r,h) = \pi r^2 h = 808.5$ cubic inches

c. We solve the equations

$$2\pi h + 2\pi r + 2\pi\left(r + \frac{9}{8}\right) = \lambda 2\pi rh$$

$$2\pi r = \lambda \pi r^2$$

$$\pi r^2 h = 808.5$$

by isolating λ in the second equation and h in the third equation to obtain
$\lambda = \frac{2\pi r}{\pi r^2} = \frac{2}{r}$ and $h = \frac{808.5}{\pi r^2}$. Substituting these expressions into the first equation gives

$$\frac{1617}{r^2} + 4\pi r + \frac{9\pi}{4} = \frac{3234}{r^2}$$

Solving for r gives $r \approx 4.87$ inches, $h \approx \frac{808.5}{\pi(4.87^2)} \approx 10.86$ inches, and

$S(r,h) \approx 519.5$ square inches.

21. The condition $\frac{f_x}{g_x} = \frac{f_y}{g_y}$ is equivalent to guaranteeing that the slope of the extreme-contour curve is the same as the slope of the constraint curve at their point of intersection.

22. It is important that the estimated point is actually on both the constraint curve and the extreme-contour curve.

Section 10.4 Least-Squares Optimization

1. a. $f(a,b) = (7 - a - b)^2 + (11 - 6a - b)^2 + (19 - 12a - b)^2$

b. $\frac{\partial f}{\partial a} = 2(7 - a - b)(-1) + 2(11 - 6a - b)(-6) + 2(19 - 12a - b)(-12) = 362a + 38b - 602$

$\frac{\partial f}{\partial b} = 2(7 - a - b)(-1) + 2(11 - 6a - b)(-1) + 2(19 - 12a - b)(-1) = 38a + 6b - 74$

The second partials are $f_{aa} = 362$, $f_{bb} = 6$, $f_{ab} = f_{ba} = 38$

$$D = \begin{vmatrix} 362 & 38 \\ 38 & 6 \end{vmatrix} = 728$$

c. Set $\frac{\partial f}{\partial a} = 0$ and $\frac{\partial f}{\partial b} = 0$, and solve the resulting system of equations.
The solution is $a \approx 1.0989$ and $b \approx 5.3736$ corresponding to an output of $f(1.0989, 5.3736) \approx 1.4066$. This is a minimum because $D > 0$ and $f_{aa} > 0$.

d. The linear model that best fits the data is $y = 1.099x + 5.374$.

3. a. $f(a,b) = (3 - b)^2 + (2 - 10a - b)^2 + (1 - 20a - b)^2$

b. $\frac{\partial f}{\partial a} = 2(2 - 10a - b)(-10) + 2(1 - 20a - b)(-20) = 1000a + 60b - 80$

$\frac{\partial f}{\partial a} = 2(3 - b)(-1) + 2(2 - 10a - b)(-1) + 2(1 - 20a - b)(-2) = 60a + 6b - 12$

The second partials are $f_{aa} = 1000$, $f_{bb} = 6$, $f_{ab} = f_{ba} = 60$.

$$D = \begin{vmatrix} 1000 & 60 \\ 60 & 6 \end{vmatrix} = 2400$$

Set $\frac{\partial f}{\partial a} = 0$ and $\frac{\partial f}{\partial b} = 0$, and solve the resulting system of equations.
The solution is $a = -0.1$ and $b = 3$, corresponding to an output of $f(-0.1, 3) = 0$.
This is a minimum because $D > 0$ and $f_{aa} > 0$.
Because the minimum SSE is zero, we know that the line of best fit is a perfect line—that is, all of the data points lie on the line.

c. The linear model that best fits the data is $y = -0.1x + 3$ percent, where x is the number of years since 1970.

d. Answers will vary depending on the year.

5. a.

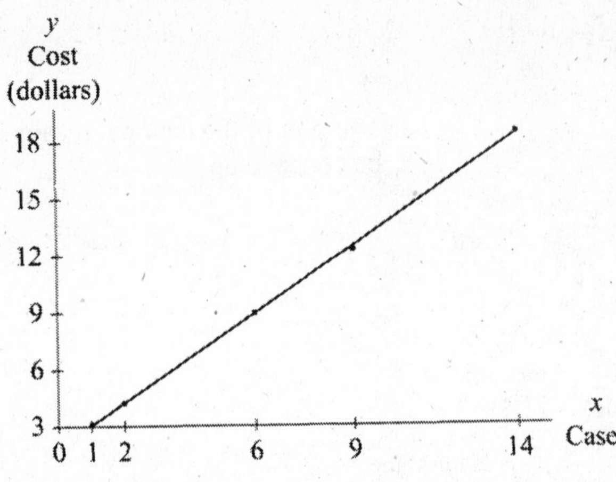

b. Using technology, $y = 1.176x + 1.880$ dollars to make x cases of ball bearings. The vertical intercept is the fixed cost per case. The slope is the cost to produce one case.

c.

x	$y(x)$	**Data value**	**Deviation: data – $y(x)$**	**Squared deviations**
1	3.0561	3.10	0.044	0.00192
2	4.2324	4.25	0.018	0.00031
6	8.9375	8.95	0.013	0.00016
9	12.4663	12.29	–0.176	0.03108
14	18.3477	18.45	0.102	0.01047
			Sum of squared deviations ≈ 0.044	

d. To find the best-fitting line, first construct the function f with inputs a and b, which represents the sum of the squared errors of the data points from the line $y = ax + b$. Find the partial derivatives of f with respect to a and b. Simplify the partials, and find the point (a, b) where the partials are simultaneously zero. These are the coefficients of the model given in part b. The function f evaluated at (a, b) gives the value of SSE shown in part c.

7. $f(a,b) = (5.5 - b)^2 + (5 - 5a - b)^2 + (4.8 - 8a - b)^2 + (4.6 - 10a - b)^2$

$$\frac{\partial f}{\partial a} = 2(5 - 5a - b)(-5) + 2(4.8 - 8a - b)(-8) + 2(4.6 - 10a - b)(-10) = 378a + 46b - 218.8$$

$$\frac{\partial f}{\partial b} = 2(5.5 - b)(-1) + 2(5 - 5a - b)(-1) + 2(4.8 - 8a - b)(-1) + 2(4.6 - 10a - b)(-1)$$

$$= 46a + 8b - 39.8$$

Set $\frac{\partial f}{\partial a} = 0$ and $\frac{\partial f}{\partial b} = 0$, and solve the resulting system of equations. The solution is $a \approx -0.0885$ and $b \approx 5.4841$.

The linear model that best fits the data is $y = -0.0885x + 5.4841$ million experiments, where x is the number of years since 1970.

9. a.

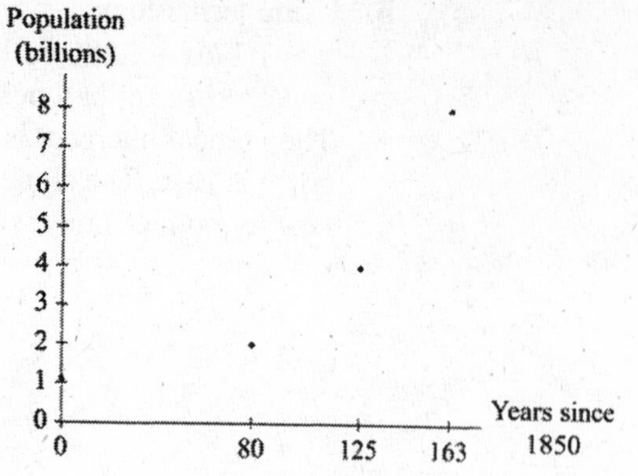

The plot of the data points appears to be concave up.

b. Plot the data points (0, ln 1.1), (80, ln 2.0), (125, ln 4.0), and (163, ln 8.0). The plot of the data points also appears to be concave up, but less so than the plot in part *a*.

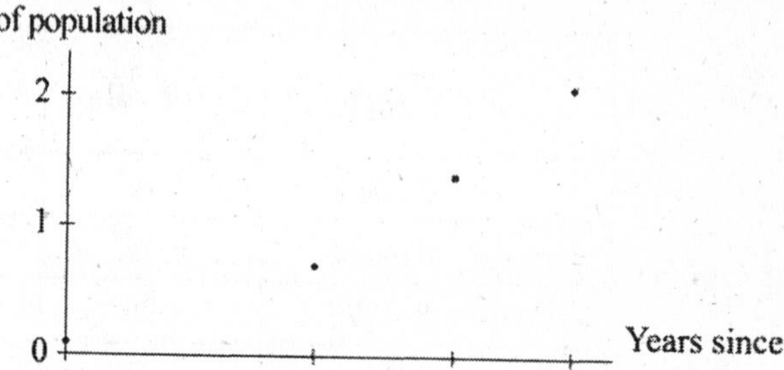

c. Using the least-squares technique, we begin with the function describing the sum of the squared errors:

$$f(a,b) = (\ln 1.1 - b)^2 + (\ln 2.0 - 80a - b)^2 + (\ln 4.0 - 125a - b)^2 + (\ln 8.0 - 163a - b)^2$$

Next we find the first partial derivatives and set them equal to zero:

$$\frac{\partial f}{\partial a} = 2(\ln 2.0 - 80a - b)(-80) + 2(\ln 4.0 - 125a - b)(-125) + 2(\ln 8.0 - 163a - b)(-163)$$

$$= 97{,}188a + 736b - (160\ln 2 + 250\ln 4 + 326\ln 8) = 0$$

$$\frac{\partial f}{\partial b} = 2(\ln 1.1 - b)(-1) + 2(\ln 2.0 - 80a - b)(-1) + 2(\ln 4.0 - 125a - b)(-1) + 2(\ln 8.0 - 163a - b)(-1)$$

$$= 736a + 8b - 2(\ln 1.1 + \ln 2 + \ln 4 + \ln 8) = 0$$

The solution to this system of linear equations is $a \approx 0.012$ and $b \approx -0.037$.

$y = 0.012x - 0.037$ whose output is the natural log of the population in billions x years after 1850

d.

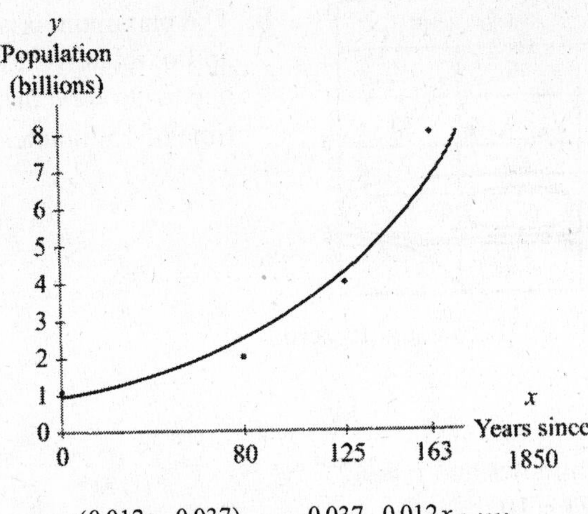

$y = e^{(0.012x-0.037)} = e^{-0.037}e^{0.012x}$ billion people x years after 1850

e. Using technology, we get $y = 0.964(1.012^x)$ billion people x years after 1850. This confirms our result in part *d* because $0.964 \approx e^{-0.037}$ and $1.012 \approx e^{0.012}$.

11. *One possible answer:* A single large outlier will not have as profound an influence on the overall fit of the line if absolute errors are used as it would if squared errors are used. However, algebraically simplifying (and solving) the sum of absolute error expressions is much more complicated than simplifying the sum of squared algebraic expressions.

Chapter 10 Concept Review

1. a.

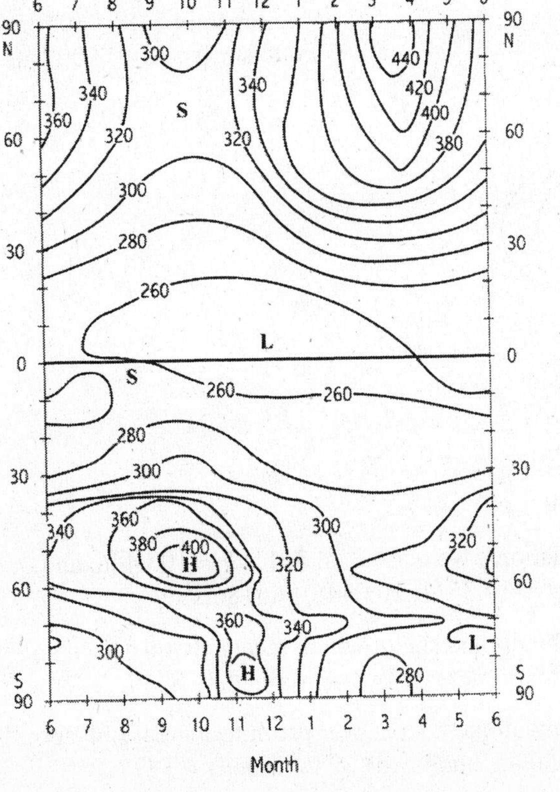

b. Highest ozone level:

Approximately 450 thousandths of a centimeter at 90°N in mid-March.

Lowest ozone level:

Approximately 250 thousandths of a centimeter at or just north of 0° (the equator) between October and March.

2. a.

Blanching temperature (°C)	Blanching time (minutes)		
	2	15	30
50	4.2	6.8	4.2
60	4.5	4.7	7.1
70	8.2	11.4	8.6
80	7.3	6.9	3.9

b. The maximum crispness is approximately 11.4, occurring for a blanching time of 15 minutes at 70°C.

c. Find the partial derivatives of C and set them equal to zero.

$C_x = -10.4x + 299.7 - 0.4y = 0$

$C_y = -0.2y + 23.1 - 0.4x = 0$

Solving for x and y gives $x \approx 26.4$ minutes and $y \approx 62.7$ °C. The crispness index is $C(26.4, 62.7) \approx 10$.

d. $C_{xx} = -10.4$, $C_{yy} = -0.2$, $C_{xy} = C_{yx} = -0.4$

$D = \begin{vmatrix} -10.4 & -0.4 \\ -0.4 & -0.2 \end{vmatrix} \approx 1.92 > 0$ and $C_{xx} < 0$, so the point is a maximum. Because the point corresponds to the only critical value and the function decreases in every direction away from the maximum, it is an absolute maximum.

3. a. The maximum profit is approximately \$202,500 for about 52,000 shirts and 9000 hats sold.

b. $4s + 1.25h = 150$

c.

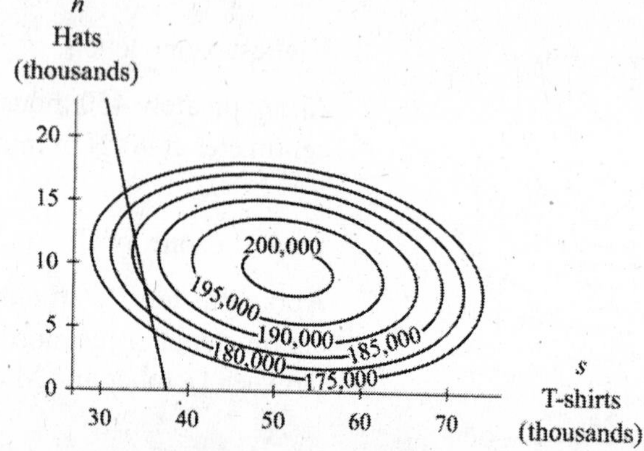

The constrained maximum revenue is approximately \$185,000 for about 34,000 shirts and 10,000 hats sold.

4. a.

$-104.4s + 5935.5 - 59.1h = 4\lambda$

$-59.1s + 10{,}299.3 - 768.6h = 1.25\lambda$

$4 + 1.25h = 150$

b. Solving this system of linear equations, we get $s \approx 34.3611$, $h \approx 10.0446$, and $\lambda \approx 438.64$. The corresponding profit is about $P(34.3611, 10.0446) \approx \$186{,}599$.

c. $\lambda \approx \$438.64$ of profit per thousand dollars spent, which means that for each additional thousand dollars budgeted, profit will increase by approximately \$439.

d. The change in budget is 2 thousand dollars, so expect profit to increase by about (\$438.64 of profit per thousand dollars budgeted)(\$2 thousand) $\approx \$877$.

5. a. $f(a,b) = (29.9 - 10a - b)^2 + (33.4 - 15a - b)^2 + (37.5 - 20a - b)^2$

b. $\frac{\partial f}{\partial a} = 2(29.9 - 10a - b)(-10) + 2(33.4 - 15a - b)(-15) + 2(37.5 - 20a - b)(-20)$

$= 1450a + 90b - 3100$

$\frac{\partial f}{\partial a} = 2(29.9 - 10a - b)(-1) + 2(33.4 - 15a - b)(-1) + 2(37.5 - 20a - b)(-1)$

$= 90a + 6b - 201.6$

Set $\frac{\partial f}{\partial a} = 0$ and $\frac{\partial f}{\partial b} = 0$, and solve the resulting system of equations. The solution is

$a = 0.76$ and $b = 22.2$, corresponding to $f(0.76, 22.2) = 0.06$.

$f_{aa} = 1450$, $f_{bb} = 6$, $f_{ab} = f_{ba} = 90$

$$D = \begin{vmatrix} 1450 & 90 \\ 90 & 6 \end{vmatrix} = 600$$

This is a minimum because $D > 0$ and $f_{aa} > 0$.

c. The linear model that best fits the data is $y = 0.76x + 22.2$ kilograms, where x is the body temperature in °C. The sum of the squared deviations from this line is 0.06, and this is the smallest possible sum.